Green Social Work

Green Social Work

From Environmental Crises to Environmental Justice

Lena Dominelli

polity

First published in 2012 by Polity Press

Polity Press
65 Bridge Street
Cambridge CB2 1UR, UK

Polity Press
350 Main Street
Malden, MA 02148, USA

ISBN-13: 978-0-7456-5400-3
ISBN-13: 978-0-7456-5401-0(pb)

A catalogue record for this book is available from the British Library.

Typeset in 10 on 12 pt Sabon
by Toppan Best-set Premedia Limited
Printed and bound by CPI Group (UK) Ltd, Croydon, CR0 4YY

The publisher has used its best endeavours to ensure that the URLs for external websites referred to in this book are correct and active at the time of going to press. However, the publisher has no responsibility for the websites and can make no guarantee that a site will remain live or that the content is or will remain appropriate.

Every effort has been made to trace all copyright holders, but if any have been inadvertently overlooked the publisher will be pleased to include any necessary credits in any subsequent reprint or edition.

For further information on Polity, visit our website: www.politybooks.com

Dedicated to my mother, Maria G Dominelli
and to my great friend, Katherine A Kendall

Contents

Detailed table of contents

Acknowledgements

There are so many people to thank when writing a book that encompasses so much territory. You are located all over the world, and too numerous to mention by name. But my gratitude for your inspiration, help and words of comfort is extensive. Without your encouragement and assistance this book would not have been written. Also, I am indebted to the Economic and Social Science Research Council (ESRC) for providing the bedrock on which Green Social Work rests. Its funding of 'Internationalizing Institutional and Professional Practices: Community Partnership Models of Change in Post-Tsunami Sri Lanka', a project on the 2004 Indian Ocean Tsunami, and the experiences of supporting disaster survivors through the Disaster Intervention on Climate Change Committees of the International Association of Schools of Social Work (IASSW) sparked my interest and research into a wide range of disasters, resilience and the indomitable human spirit.

I want to thank my family for the endless support they have offered when I am stressed out by the writing project, technology going wrong, and all the other awful things that happen in daily life that call upon our resilience and supportive networks. You were always there for me and helped me move on with the task to hand.

I also thank the publishers for their endless words of encouragement and willingness to see the project to its completion. And, although I express my gratitude to all my colleagues at the School of Applied Social Sciences and the Institute of Hazards, Risk and Resilience Research at Durham University for their support of this venture, I wish, however, to identify and especially thank several of them individually – Tom McLeish, Stuart Lane and Simon Hackett – for ensuring that I engaged wholeheartedly with the physical and natural scientists to show that the world can become a better place if the disciplinary divide in the sciences can be overcome and we work together.

Lena Dominelli

1

Introduction

Setting the Scene

Environmental crises and their impact on diverse populations across the globe have challenged social work practice at the beginning of the second decade of the twenty-first century not only by the frequency of their occurrence, but also by their complexity and the substantial damage caused to the Earth's physical environment and well-being of the countless numbers of people, animals and plants living in it. Those linked to natural hazards such as the 2010 earthquakes that occurred in Haiti, Chile and Christchurch and the 2011 earthquakes in Christchurch and Japan, the latter of which also encompassed multiple hazards caused by a tsunami and radioactive leakages, have overwhelmed the helping services by the magnitude of the damage caused. They have also highlighted the linkages between human behaviours, economic imperatives and social policies and the extent of suffering experienced by victim-survivors. For example, corruption among the elites of Haiti, the inability to observe building regulations in Chile, and corporate secrecy in Japan meant that the damage was more extensive than it need have been and that emergency responses were hampered by the lack of information, infrastructures and resources vital for dealing with these catastrophes.

Poor women, children and men bear the brunt of these failures. Victim-survivors' experiences demand transformations in the conceptualization of and responses to disasters, as does caring for the environment. Social workers, as the professionals responsible for safeguarding human well-being must rise to the challenges posed by this state of affairs.

While social workers have provided humanitarian assistance and counselling services to those on the ground, their voice has scarcely been heard in the media. Additionally, it is absent from many of the decision-making structures formulating policy for preventing large-scale devastation in the future, and addressing needs during calamitous events and afterwards. Social workers' lack of involvement is not unexpected. Their tasks involve supporting those requiring assistance. But they rarely have time to engage with the wider issues of practice, including the development of social policies that draw lessons from micro-level practice and affirmation of their expertise when other professionals extending their roles seek to appropriate it. The profession has a wide-ranging remit, but social workers have played a minor role in deliberations about the deleterious impact of environmental disasters on people's well-being, in local communities and globally.

Except for community workers, those intervening in deprived localities with poor housing stock have seldom taken action to raise consciousness about the appalling physical surroundings in which service users or clients live. Social workers might advocate for a family needing a new home to protect children suffering from respiratory ailments caused by damp, mouldy housing, but, unless they are community workers, they would not engage in collective action to repair or replace decrepit properties in an entire estate. Their voice has been virtually absent from climate change discussions. Yet, they work in different communities in the aftermath of environmental disasters, whether these are caused by climate change, nature, or industrial accidents like the one that occurred in Bhopal, India, or conflicts between populations like those happening in the slums of Nairobi in Kenya. In the latter, refugees escaping the loss of grazing lands through droughts that had destroyed traditional life-styles became embroiled in tense confrontations with existing slum dwellers.

This book contributes to filling the gap in the literature caused by the shortage of publications that specifically address environmental issues from a social work perspective, advocating for and strengthening the voice of social workers who support people during disasters at policy-making and practice levels, however and wherever these take place. There are a limited number of texts that address what has been termed environmental social work and ecological social work. I focus on what I call *green social work* because I want to produce a book that transcends the concerns of ecological social work, which is a systems-based approach to the mainstream social work preoccupation with the person in their environment, usually defined as their social environment (Van Wormer and Besthorn, 2011), e.g., Gill and Jack (2003). Such writings tend to ignore power relations based on geo-political social structures that have a deleterious impact upon the quality of life of poor and marginalized populations and the Earth's flora and fauna. They also fail to endorse action that could secure the changes necessary for enhancing the well-being of both human beings and the planet, I argue that green social work focuses on how responses to environmental crises must both challenge and address poverty, structural inequalities, socio-economic disparities, industrialization processes, consumption patterns, diverse contexts, global interdependencies and limited natural resources.

Given the profession's embeddedness in life-enhancing micro-practice in everyday routines, I argue that contemporary social work has a vested interest in attending to environmental issues as an integral part of its daily remit if it is to retain its currency in contemporary societies, emphasize its relevance to the social issues that peoples have to resolve in the twenty-first century and widen its scope if it is to prevent the haemorrhaging of its activities to related professions including health, geography, psychology and psychiatry. Additionally, practice has to engage with both local and global contexts to develop those that are locality-specific and culturally relevant and that engage with global interdependencies within and between countries. This goal configures the environment as a socially constructed meaningful discursive space that encompasses physical and material realities, socio-economic, political and cultural structures, and spiritual and emotional places that come together in one whole as the space in which individuals breathe their lives symbolically, in real and imaginary time.

I suggest that poverty is a constant, on-going disaster in its own right and not simply an additional factor to be considered in determining individual vulnerability to disasters. This sense of it is not adequately conceptualized in the existing literature. I connect this wider, structural notion of poverty to social justice claims that contemporary citizenship is denied to low-income people who cannot overcome poverty or participate in market-based solutions to social problems, including climate change. Nor can these groups afford to purchase new, renewable 'green' technologies that will enable individuals to procure energy-saving devices. In exploring these issues, I examine the interests of multinational corporations, material consumerism, the unequal distribution of resources, and population movements that are undermining attempts to conserve energy and reduce the exhaustion of natural resources, ranging from land to fuels, from minerals to air. And I link the resolution of these matters to community initiatives that social workers can engage in to ensure that the quality of life of poor people can be enhanced today and tomorrow without costing the Earth.

This strategy critiques the lack of resources associated with the cutbacks in public expenditures that are occurring throughout Western Europe, Canada and the USA as their rulers' search for solutions in the community, which they create as a responsible discursive space that assumes unlimited goodwill and regenerative self-help activities to cover the gap between what people need throughout the life course and what they can secure without state support or employment in well-paid, life-affirming jobs. The state attempts to wash its hands of its caring responsibilities. Now described as burdens, these are epitomised in the 'Big Society' ideas promoted by David Cameron in the UK. I focus on this particular example because it is instructive in exposing the sophisticated structure of the fig-leaf it represents by drawing upon society's potential to bring people together in self-help initiatives that might be useful in creating self-sufficient communities but hide the state's role in creating poverty, its indifference to people's personal suffering and its failure to control industrial barons and the financial sector and to create life-affirming and sustainable environments for people, plants and animals. The failure of states to govern for the entirety of their population, the planet's flora and fauna, and the material worlds that fall within their boundaries, exposes elite governmentality and loss of good governance that extends beyond the 'failed' states identified a few years ago by George W. Bush as President of the USA, and reveals a democratic deficit of the highest order that leaves citizens feeling disenfranchised, alienated and isolated from their rulers.

Moreover, I examine the formation of energy-sufficient communities that utilize self-help to access the new renewable technologies and reduce reliance on fossil fuels. In implementing such initiatives, community social workers can draw lessons from the skills of mobilization that middle-class communities display in creating more humane urban environments for transferral where relevant in mobilizing working-class endeavours. Such skills can be acquired without condoning deficit models of poor communities or suggesting that they eschew their own identities and adopt middle-class lifestyles as often occurs in mainstream social work (Callahan et al., 2000). Additionally, I argue that social services must become an inclusive, universal service, not a residual and stigmatized one. High-quality social services for all, readily available and accessible at the point of need, are a human right and integral to realizing social justice claims to resources. In this book, I utilize case studies to integrate theory and practice around these concerns and indicate the importance of thinking holistically about these issues.

Locating a new discursive space

Social work educators have written under the rubric of ecological or environmental social work. Their works include Van Wormer and Besthorn's (2011) *Human Behaviour and the Social Environment*; John Coates' (2003) *Ecology and Social Work*; and Gill and Jack's *Children and Families in Context*. These primarily reflect a systems approach to the social environment and an individual's place within it. They are less concerned with structural analyses of the social and economic developments that have destroyed both physical and human socio-cultural environments. They are not embedded in the social and physical elements of life for the purposes of changing existing socio-economic relations. There is a Global Alliance for Deep Ecological Social Work that includes authors such as Besthorn and Coates which has sought to enhance social workers' interest in the physical environment. Its impact has been limited, but it has organized several conferences to discuss these issues. Michael Ungar's (2002) article on environmental rights for marginalized groups and the formation of sustainable communities brings a deeper dimension into these debates by highlighting the importance of human rights. Although he addresses social work issues, his work lacks the sweep of this particular book. Nor does Ungar utilize case studies that could encourage practitioners to work outside their usual boxes in bringing about a more sustainable, and socially and environmentally just, world.

My arguments also go beyond relevant works like Cahill and Fitzpatrick's (2002) edited book, *Environmental Issues and Social Welfare* which is not about social work practice as such. None of the authors in their edited collection is engaged in mainstream social work; nor do they cover practice. Another one, *Right Relationship: Building a Whole Earth Economy*, written by Brown and Garver (2009), helpfully critiques capitalist economic development particularly that of neoliberal economics, for destroying human well-being. But it has little to say about practice issues, including those of organizing residents to resist neoliberalism's destructive tenets and argue for more collective and interdependent approaches to the economy, as social workers do when supporting people in rebuilding their lives through alternative economic community models, many of which draw upon micro-finance and credit union initiatives that pool resources collectively. Kati Närhi (2004) considers structural inequalities in what she terms the 'eco-social approach in social work', but concentrates on a narrow range of ecological issues and a small area in Finland. The objectives in Nancy Mary's (2008) book are closer to mine, but these do not cover practice in a sustainable world, nor a critique of industrialization and consumerism as these influence practice interventions, and so has gaps that I consider. McKinnon (2008) highlights the relevance of the value base of social work in addressing ecological issues, while having a limited engagement with environmental matters. Given its scope and range, *Green Social Work* aims to break new ground.

This book engages with environmental and social justice issues in the contexts of eradicating structural inequalities by critiquing an industrialization model that caters for the needs of the few, in order to enhance humanity's and the planet's well-being. It is also embedded in a collective duty to care for, and be cared by, others. Being green in social work encapsulates a holistic approach that addresses both personal behaviour and the structural facets of social organization and marginality to argue for mutuality and solidarity in solving social problems that are rooted in an unequal distribution of: the Earth's resources; its technological innovations; and social provisions that can be employed to enhance human well-being. These have to be spread equitably across the globe while at the same time acknowledging interdependencies between peoples and other living things, and showing respect for the Earth's limited physical resources, its flora and fauna. The challenge for green social workers is that of working to enhance the quality of life for marginalized peoples today while also preserving the Earth's largesse for future generations. Embedding this in looking after the well-being of all individuals, animals and plants is crucial in developing alternatives to existing unequal social relations. Enhancing planetary well-being requires a diminution of consumerist lifestyles as typified by Western societies and burgeoning middle classes in emerging economies like India, China and Brazil; the development of new economic paradigms that take account of the needs of all stakeholders, not just those who want to make a profit from their investments; greater planning for projected growth in human populations; and the protection of a diverse biosphere and physical landscape.

Green Social Work also seeks to create a specific subject of study in an area that has been largely neglected in social work by focusing on the interaction between equality, securing the well-being of people, animals and plants, and protecting their physical environments including the built environment that covers housing stock, power grids, transportation and communications infrastructures and the natural environment, including land, air, water, mineral resources and primary products within the context of environmental rights and social justice. The social and environmental justice dimensions of this topic bring marginalization, structural inequalities, human rights and active citizenship into the heart of the green social work agenda and call for the creation of new models of intervention within a framework of preserving Planet Earth. To achieve this, rethinking neoliberal capitalist relationships between peoples and their environments becomes unavoidable.

The rationale for green social work

Social work as a profession has engaged in environmental issues and continues to do so, albeit in a limited rather than comprehensive manner. The intermediate treatment programmes extremely popular in the 1980s in the UK, which involved outdoor activities for diverting young offenders from prison, exemplify this. They encouraged personal growth in young people, and taught them skills in relating to other people and the physical environment. These endeavours declined in popularity when the tabloid press attacked them for being 'jollies' at the taxpayers' expense and the state responded by refusing to fund alternatives to prison like these. For example, in England, Mark Hook, who had a string of offences to his name, was dubbed 'Safari Boy' by the media after going on an 88-day character-building trip to Africa at the cost of £7,000. Soon after his return, he was imprisoned for nine months by a Gloucestershire County Court for burgling a house, stealing several items and driving a car without insurance after absconding from the Bryn Melyn Centre in North Wales (Waterhouse, 1994). The media hysteria around 'Safari Boy' provided a crucial 'tipping point' in the discourses about intermediate treatment because young offenders like him committed further crimes after completing such programmes. Their reoffending undermined the programme's claim to change individual behaviour by building character and facilitating personal growth. By focusing on the young person's failures, the media and state conveniently ignored state failure in changing the lack of opportunities and conditions of deprivation prevailing in the communities to which young offenders were returned. Personal and structural changes have to occur together to support successful interventions in people's lives. This is a more constructive lesson to be drawn from this affair.

Green social work, as I define it, has holistic understandings about the various environments and their impact upon people's behaviour. Although utilized for therapeutic purposes, the environment is a socially constructed entity in and of itself, not a means to an end and should be respected as such, even when people use it to meet their own goals. I suggest that the failure in probation officers' involvement with 'Safari Boy' was in expecting a short outdoor experience in unknown territory to achieve their objectives of controlling and disciplining a young man rather than inculcating a sense of valuing himself and the environments that he already knew on a day-to-day basis, and understanding how these blocked his ability to realize his full potential and what needed to be changed in these, as well as in his behaviour, to secure enduring changes that would facilitate his becoming an active, valued member of society. Had they done so, this might have produced the self-confidence and trust in himself, others and the environment needed for real behavioural change to occur. People need a sense of place and stake in society to respect and value the environment and others. Otherwise, alienation and a distancing of themselves from others will illicit further offending behaviour or social disorder. This lesson has relevance for what should happen to those sentenced for their participation in the 2011 summer riots in England or imprisoned for opposing the cuts in Greece.

I define 'green' social work as that part of practice that intervenes to protect the environment and enhance people's well-being by integrating the interdependencies between people and their socio-cultural, economic and physical environments, and among peoples within an egalitarian framework that addresses prevailing structural inequalities and unequal distribution of power and resources. Paying attention to these requires social workers to address the politics of identity and redistribution and not to treat the environment as a means to be exploited for people's ends. By being concerned with the politics of identity and the politics of redistribution, this book goes beyond the issues raised by the 'deep ecological approach' to social work, which focuses largely on the interaction between people and social and physical environments while viewing the latter as objects on which human beings act. I suggest that this continues to privilege people as separate objects and does not integrate all environments – physical, social, economic, political and cultural – within which people are embedded into a holistic social work practice that engages with and aims to change existing inegalitarian social relationships, power relations and resource distribution systems. In this sense, the social and physical environments are inter-related and interact with and impact upon each other. As I consider these elements, I examine the roles that social workers have played and can play in the key environmental issues of our time – environmental degradation; industrial pollution; over-consumption by the few; climate change; migrations caused by natural disasters; and conflicts that will increasingly accompany peoples' competition for scarce natural resources like water, land and clean air. This includes exploring social workers' capacity to act as: advocates; mobilizers and organizers of people, communities and resources; lobbyists who can influence policy-makers; and therapeutic workers who can respond to individual distress. I then consider how social work practice can be transformed by encompassing a holistic 'green' agenda rooted in the interdependency of all peoples and their socio-cultural, economic and physical environments. I conduct my arguments as indicated in the chapter outlines below.

The Structure of the Book

In chapter 2, I consider how poverty, a key social disaster, is accompanied by the lack of environmental rights, with poor people living in the most degraded social and physical environments and disproportionately subjected to industrial pollution and natural disasters. Poor people lack the financial wherewithal to strengthen their capacity to cope with environmental crises or to purchase expensive fuel, high-quality food and decent housing. The current global fiscal crisis is decreasing public welfare provisions, particularly those upon which poor people rely as demand for them is rising. Undermining a nation's capacity to repay debts by letting the markets decide whether a sovereign state can be considered creditworthy enough to do so, whilst compelling its peoples to endure ever tighter austerity measures, can result in public funds lining the pockets of hedge fund holders, as is anticipated to occur over Greek bonds, an example of predatory capitalism that is likely to incur taxpayers' wrath when they contrast speculators' enormous profits with their own penury (Landon, 2011). In other Western countries, the 'age of austerity' prevails, but imperils Western economies because they cannot produce the growth that is needed to promote prosperity and take people into paid employment. President Barack Obama is struggling to address entrenched unemployment in the USA as the economy slides into deeper recession. In the UK, Prime Minister David Cameron proposes grandiose ideas like the 'Big Society' to get people through hard times. This scheme exacerbates poor people's plight by combining savage public expenditure cuts and reduced publicly funded welfare resources with pathologizing people, blaming them for their predicament, and emphasizing their lack of skills, initiative and will-power to pull themselves up by their bootstraps. Contextualized, holistically embedded, innovative strengths-based solutions will be necessary for people to respond effectively to such onslaughts. Social workers can assist them in finding new paths forward. I examine such initiatives in chapter 2 in light of social workers' roles within both social and physical environments.

I also consider the crisis in the profession itself as it endeavours to prevent the haemorrhaging of its borders to other professions, particularly psychology, psychiatry, human geography and criminology. At the same time, social workers have opportunities to expand into new arenas by addressing environmental issues and practising green social work within a redistributive framework that operates locally and globally.

Social workers have the skills to deal with new challenges, as they have done when community workers have mobilized communities to protect them from the predations of the market. These are instanced by resistance to the ghetto cleansing that occurs when poor people in the West resist being removed from their homes to make way for a major road, or poor people living in urban slums in Mumbai, India, refuse to cede their homes to prestige developments for wealthy people. Social workers help poor people mobilize as groups to challenge such developments. Some of these struggles have been won, others have been lost. An important point, however, is that social workers continue to support poor people who seek to defend their communities and interests, even when powerful decision-makers and resource-holders defy their requests for better-quality living conditions.

The central theme of chapter 3 is industrialization and urbanization, in the context of a critique of its current forms. Industrialization has centralized activities in urban environments while built-infrastructures have opened up opportunities for business to make profits and encourage peoples' mobility in search of jobs and improved standards of living. This approach has led to hyper-urbanization and enormous damaging consequences for peoples' well-being and the physical environment. Although poor people bear the direct costs of living in degraded physical environments, they seldom have a say in how development initiatives are executed and what impact these will have on them and their communities. This has been evidenced historically during the Enclosures that left rural people landless in the Scottish and English countryside and subjected countless inhabitants to destitution in the big cities of Victorian England, e.g., London. It is currently evident in shanty-towns in megacities of the Global South. The outcome in both scenarios has been the same: a drop in the quality of life for once self-sufficient rural peoples in the short term, and a dependency on their ability to sell their labour in the long term. The lack of sustainable, healthy environments and well-paid employment opportunities in rural areas must be addressed to limit the pressures and problems of hyper-urbanization.

Social relations also became more hierarchical and differentiated, within the home where men began to dominate, and in the workplace where employers called the tune that those fortunate enough to have paid work were expected to obey. Under this form of social organization, it has been difficult to hold those with wealth accountable for their behaviour and the decisions they make, despite their negative impact on other peoples' livelihoods and existence. Lack of accountability among the owners and managers of contemporary multinational corporations remains a problem that social workers can highlight. Issues have to be identified and named before they can be addressed. Practitioners can help

communities organize to hold such firms accountable for the consequences of their decisions on the lives of usually voiceless people.

In chapter 3, I use case studies to consider how social workers can
operate at the local level to develop more sustainable and life enhancing
forms of urban living and demand accountability for actions taken by
all stakeholders involved in local areas. I explore how local schemes can
have an impact at national and international levels and provide insights
for improving situations through collective action. I draw on examples
from both the Global North and the Global South to examine these
issues. These include: the formation of micro-credit ventures; locally
accountable financial institutions such as credit unions; creation of local
area networks such as local exchange schemes, to avoid monetary transactions while providing people with access to services outside the market-
place; and formation of social enterprises. I adopt a critical reflective
stance towards such initiatives, because social workers have much to
learn about intervening in these activities holistically.

Industrialization continues as a key topic for chapter 4. This time I
focus on how pollution, as a by-product of industrialization, has had
deleterious outcomes for people's health, e.g., increasing respiratory ailments and a range of disabilities that can be traced back to the lack of
controls on the pollutants that industrial processes discharge into the
atmosphere. The banning of lead as an ingredient in petrol-driven cars
in the West illustrates how changes in daily routines can alter everyday
behaviours, like driving, to improve significantly the health of people
living near major roads. Serious accidents have resulted from human
control over scientific products going awry. The nuclear explosion in
reactor 4 in Chernobyl in 1986 exemplifies such a problem. It entailed
wide-ranging and long-term consequences that continue to be addressed.
Other examples have been the leakage of dioxins or radioactive materials
in many spots throughout the world, including the discharge of dioxins
into the atmosphere in Seveso, Italy, in 1976, and the more recent nuclear
explosion following an earthquake and subsequent tsunami in Japan's
Fukushima Daiichi nuclear power plant in 2011. Such accidents have
gravely undermined people's well-being, defined as their right to enjoy
the products of the Earth and develop their skills and talents to the fullest
extent as articulated under Articles 22 to 27 of the Universal Declaration
of Human Rights (UDHR).

I examine the huge industrial accidents at Three Mile Island in the
USA, Chernobyl in the Ukraine, and Bhopal in India from a social work
perspective in chapter 4. Poor health and higher morbidity arising from
increased rates of cancer, higher levels of congenital disabilities, loss of
livelihoods, social isolation and stigma are concerns that social workers
have addressed. I also highlight the importance of utilizing existing social

resources and legal instruments to hold perpetrators accountable for their actions, while exploring social workers' roles in developing local resilience in response to industrial disasters like these.

Climate change is the major theme of chapter 5. Climate change is the by-product of industrialization models promoted by Western entrepreneurs exploiting natural resources to produce goods that make a profit, while discharging greenhouse gases and other pollutants into the atmosphere and water because this was the cheapest method of waste disposal. These activities subsequently caused air temperatures to rise to levels that threaten all forms of life on Earth. Climate change portrays recent environmental crises that embody global interdependencies whereby what one country does can have severe and damaging impacts upon other nations and their environments. The benefits that the West has gained arose through industrialization processes that have increased greenhouse gas emissions to the detriment of people living less industrialized lifestyles, such as poor people living in rural areas of the Global South. These outcomes are especially worrying in small island nations that are in danger of sinking into the ocean, e.g., the Maldives, Tuvalu. Although unintended, these consequences have resulted in claims for the 'polluters to pay' for cleaning up the environment and developing renewable green energy sources that will limit temperature increases to less than 2 °C in perpetuity.

Regardless of where they live, people without incomes or on low ones cannot access the expensive technologies that would enable them to protect, maintain or enhance their quality of life without state intervention and support. I argue for states to play an important role in this because: (a) the nation-state, as the embodiment of the collective will of the people and guarantor of their rights to be cared by, and care for, others, has to be held accountable for its failure to deliver these for all its residents; (b) the market as currently constituted cannot fulfil this function for all; and (c) charitable giving is insufficient for the enormity of the tasks to be addressed. Moreover, the state has to engage corporations as philanthropic players in solving these problems. Politicians can express their choices by subsidizing private firms and individuals; building public infrastructures that will sustain renewable energy sources; and public education campaigns to initiate behavioural change at the individual level. Current strategies of 'business as usual' do not challenge the industrial model that is responsible for creating these problems in the first place. Nor do they begin to hold accountable the industrial entrepreneurs who ignore the consequences of their decisions for those least likely to have the resources to tackle them. These messages have been repeated on many occasions at the meetings on climate change hosted by the United Nations. So, states must rethink their priorities and responses. The interdependencies between countries, peoples and the environment suggest that solutions to climate change dilemmas have to be more inclusive and integrate those who are excluded from the market-place of expensive technologies in finding solutions to controlling energy consumption, and the problems raised by an unequal distribution of power and resources globally. Social workers can both highlight such problems and engage with local people in responding to them.

Poor people's capacity to transcend financial limitations in energy reduction strategies might be an outcome of the proposed 'feed-in' tariffs in the UK. In this scheme, residents who create electricity for domestic use through renewable energy sources and 'feed' the units surplus to their needs into the national grid are paid a fee set by government for these extra units – hence the term 'feed-in tariffs'. Whether their introduction will benefit poor peoples and their communities as utility companies form partnerships with them remains to be seen. However, there are small demonstration projects tackling issues like fuel poverty and unemployment through the development of micro-renewable energy technologies which will be considered in this book. By engaging in such initiatives, poor people can address fuel poverty, reduce fossil fuel energy consumption and develop self-sustaining energy communities while enhancing employment opportunities and raising their standard of living. I consider such endeavours in this chapter because community social workers have been involved in these in both the Global North and the Global South, e.g., Gilesgate in the UK, Misa Rumi in Argentina. I use such examples to argue that social workers can play an active role in climate change policy debates initiated under the Kyoto Protocols and expand the domains in which they are proactive.

Environmental degradation has damaged physical environments to produce desertification, flooding and other environmental crises. In chapter 6, I consider the consequences of climate change and industrial processes that encroach into traditional farmlands and forests. These have exacerbated the loss of traditional habitats and lifestyles, particularly nomadic ones in the Global South and other countries that have indigenous people who favour traditional ways of life that link people and land holistically. Environmental degradation has led to conflicts between groups of people who have migrated in response to deteriorating conditions in their surroundings. The ensuing migrations, in certain parts of the world subjected to either desertification or extensive and prolonged flooding, have intensified land stress in both rural areas and built environments in cities, as populations compete for scarce social and physical resources and spaces. Competition between people and scarce resources on the land can also endanger people and their surroundings in refugee camps created to house those fleeing environmental crises.

Building resilience in such communities to address pre-disaster and post-disaster situations is an important part of reducing vulnerability to climate change. In this chapter, I consider how social workers act as mediators in conflict situations, and as development workers to help rebuild lives and communities in more sustainable ways. An interesting example of social workers' interventions occurred in the Mathare Valley in Kenya where practitioners worked with residents to bring harmony to a volatile situation involving newcomers and more established dwellers when nomadic peoples migrated to the slums of Nairobi to escape drought.

Other problems explored in chapter 6 include those of tribal peoples living along the borders of Kenya, Somalia and Ethiopia who were prevented from moving across borders by boundaries formed during the European colonization of Africa. For example, Dadaab camp in Kenya was created for 90,000 Somalian refugees following the collapse of the Said Barre dictatorship. The 2008–9 drought more than doubled the numbers descending upon this camp from these three countries, thereby causing further environmental degradation. Official attempts to limit the growth of these settlements included one whereby the Kenyan government declared that, if its nationals sought support at the camp, they would lose their citizenship rights. This form of exclusion failed to stop the Masaii peoples from going there. This issue raises concerns about the Kenyan state's failure to uphold its citizens' rights under the UDHR. Furthermore, social workers as relief workers staff refugee camps for any displaced population and would deem it unethical to uphold any stipulation that excludes any particular nationality from receiving assistance if they arrived at Dadaab. The plight of the Masaii also raises the thorny issue of the arbitrariness of the frontiers that colonial powers imposed on local people in Africa. And it poses the question of why people as citizens, should lose their rights to ask for help from their national governments simply because they cross into lands that they had traditionally used as herds-people. Moreover, food provided to refugee camps through the World Food Programme is usually insufficient, but aid workers are dependent on government and the public for the resources they need to do this job. Food shortages occurred elsewhere during this drought, e.g., in the camp at Ayub in Ethiopia. In considering these matters, I explore other roles that social workers can play in these circumstances, e.g., questioning the loss of citizenship rights to state assistance.

Environmental degradation brought about by industrialization, urbanization and the demands of growing numbers of people on Planet Earth have brought about failures in the built environment and infrastructures that sustain peoples, to the detriment of their livelihoods and well-being when 'natural' disasters strike. I address these in chapter 7, by exploring the complex connections between population movements, marginalization and social exclusion and demands on environmental resources. These competing requirements can be addressed with sensitive planning and sufficient community engagement. To achieve this goal, humanitarian aid workers cannot work according to a 'one size fits all' plan. Instead, they have to contextualize their work, engage effectively in locality-specific and culturally relevant practices with a range of different agencies, academic disciplines and government bodies that support poor people.

The impact of Hurricane Katrina on New Orleans is a recent example of the havoc that can occur when relief and government responses are unable to meet the needs of poor and marginalized peoples, in this case, of African Americans. The Japanese government's early responses to Japan's multiple hazards in 2011 were also declared inadequate by victim-survivors, especially for older people. Both examples indicate the paucity of responses even though these involved two of the richest countries in the world. In Katrina, that people of African American origins, older people, and low-income families with children were disproportionately adversely affected during relief efforts after the levees broke, demonstrated how those most marginalized by society suffer most during relief efforts, as well as through the disaster itself. The floods in Pakistan during the summer of 2010, and the Haitian earthquake in the same year, were followed by cholera outbreaks and gastrointestinal disease. These two situations illustrate the lack of urgency in rebuilding infrastructures, including sanitation and water supplies when a disaster destroys these. And they expose the failure of global responses to meet the needs of poor people.

In chapter 7, I also examine social workers' involvements in calamitous 'natural' disasters, practitioners' critiques of disaster interventions and their suggestions for providing more appropriate ones in the future. I argue that 'natural' disasters have a substantial (hu)man contribution that exacerbates their deleterious impact. I also suggest that reactions to such events can be more preventative in their focus if they build resilience among peoples and communities for dealing with such events and rebuilding their lives afterwards. The involvement of communities as full players in these plans is vital to enhancing resilience at all levels, and is an element of practice that social workers can advocate for and promote. I also endeavour to examine how interdisciplinary teams of experts can work closely with local people to prepare better for such events in future. For example, in Haiti, landslide experts can work with relief workers to ensure that refugee camps are located in safe spaces.

In chapter 8, I argue that the Earth's natural resources, such as land, water, energy supplies and minerals, are being exhausted by the demands of Western models of industrialization. These are clearly unsustainable for the few who benefit from them now, let alone the growing numbers of the world's population, which the United Nations predicts will exceed 9 billion by 2050. I consider these concerns and those caused by growing population numbers in the context of scarcity in this chapter. Whilst I do not endorse Malthusian gloom over this issue, unless ways of raising people out of poverty, promoting sustainable development and healthy lifestyles are found, the future could be very bleak for generations to come and the Earth's flora and fauna. And it could intensify conflicts over scarce resources, particularly land and water, in future.

On an optimistic note, some countries have sought to achieve non-violent means of resolving conflict over natural resources. A current example is the talks between Egypt and Ethiopia over the management of the Nile to meet the water needs of growing populations and desire to improve their standards of living in both countries. Their cooperation is in contrast to the violence that erupted in 2008 over the control of the waters flowing in the Isfara river at the border between Kyrgyzstan and Tajikistan. In chapter 8, I also examine how social workers can support inter-country initiatives aimed at resolving potential conflicts over scarce resources to achieve win-win situations for all, including Planet Earth. I add to this a consideration of policy initiatives locally, nationally, region-ally and internationally.

Chapter 9 explores modernity or Western ways of thinking about the world, based upon industrialization processes that have focused on hier-archical and binary dyads that prioritize monied people and their inter-ests. Until its critique by post-modernists, modernity as a worldview was deemed superior to others. It gloried in being the product of rational thought processes rooted in empirical evidence that its adherents called 'scientific', while disparaging others as inferior to it. The disparagement of alternative ways of viewing the world was practised with particular vehemence against indigenous or pre-industrial ways of being and living on Planet Earth. The West has benefited from such depictions of reality by raising its people out of the forms of deprivation typical in nineteenth-century Europe. But it failed to eradicate poverty within its own borders, and intensified poverty elsewhere by destroying traditional lifestyles and promoting economic underdevelopment in the interests of commanding the Earth's natural resources for its own projects. Questions are now being raised about the dangerous potential of such colonizing ventures and their capacity to be perpetuated in neo-colonial guises through edu-cation that advances Western models of thought. Concern is also being expressed about rising superpowers like China that acquire land and other resources without paying attention to local aspirations or safe-guarding the human rights of local peoples and their environments.

Alternative worldviews have existed before and continue to be available. Significant among these have been spiritual approaches to life evident among indigenous peoples in the West and elsewhere. They have struggled to keep alive their cultures, languages and traditions despite the onslaughts legitimated through colonization. Indigenous peoples in Asia, Latin America and Africa have sustained their traditional lifestyles against the odds. Like their counterparts in the West, they demand the restoration of rights over their resources and more sustainable lifestyles. These are often defined as indigenous movements, which have had a significant impact upon social work itself (Gray et al., 2008). Key among indigenous approaches to daily living are the integrated relationships that they envisage among themselves as people, and between them and their social and physical environments. Their conceptualizations of themselves as keepers of the Earth's flora, fauna and natural resources for future generations are embedded in a spiritual connection between people, other forms of life, and inanimate objects. This is referred to as a spiritual orientation to living things and their physical environments (Green and Thomas, 2007).

In chapter 9, I consider how indigenous beliefs, particularly those of First Nations in Canada and Maori in Aotearoa/New Zealand tend to be collective, aimed at causing least disruption to natural environments and leading sustainable lifestyles based in local communities. Indigenous peoples underpin their activities with more holistic and collective approaches to life, and link the individual to their specific community. They also enable the continuation of heritage languages and traditional identities. Drawing on these assets, indigenous peoples have critiqued dominant approaches to welfare and the criminal justice system to develop alternative models of dealing with those experiencing hardship and/or social problems (Grande, 2004), e.g., Maori approaches to young offenders. Practised through family group conferences, it involves the entire extended family in rehabilitating and supporting the young persons concerned to reorient their lives.

Insights derived from indigenous knowledge and approaches to life can provide lessons that could help social workers practising in densely populated urban areas, by assisting people in reconnecting with the physical world and renewing their ties with other peoples, including those in rural areas seeking to industrialize in sustainable ways that do not cause the mass migration of young people to urban centres in large cities to earn livelihoods that are difficult to come by and cause further destitution and environmental degradation.

I conclude this book in chapter 10 by considering how the contributions of previous chapters can develop a holistic model of social work practice that promotes interdependence and solidarity among the world's peoples, its flora, fauna and natural resources. This approach, promoted in and through community-based practices that hold governments and multinational companies accountable for their decisions, is crucial in developing the sustainable lifestyles that will protect the well-being of all peoples and preserve and/or enhance the environment for current and future generations. I call this model *Green Social Work*. It involves social workers working closely with people in their communities through every-day life practices to: respect all living things alongside their socio-cultural and physical environments; embed economic activities, including those aimed at alleviating poverty, in the 'social'; and promote social justice and environmental justice. This will require social workers to engage in action at the local, national, regional and international levels and to use the organizations that they have formed to advocate for changes that favour the equal distribution of power and social resources, and protection of the Earth's physical bounty, including its flora and fauna.

2

A Professional Crisis within Social and Environmental Calamities

Introduction

Social work in the early twenty-first century is facing past challenges and new ones. Persistent ones like failing to tackle poverty, juvenile and adult offending behaviours and preventing child abuse have occurred throughout the profession's existence. Others are relatively recent in its repertoire, for example, climate change and disaster interventions. At the same time, social work is undergoing considerable change as a professional discipline. Its status among high-ranking professions like medicine and law remains in doubt in both the academy and the field, despite a century of endeavours aimed at raising its research base and profile, nationally and internationally. Although a historical concern, this uncertainty has dented the profession's confidence as its energies are consumed by proving its credentials and demonstrating that it has its own specific knowledge base, and particular theories and methods of practice. Although its varied professional patterns are recognized throughout the world, social work practitioners are more valued in some countries than others. In Europe, British politicians are particularly unappreciative of the contributions that social workers make in society generally, while those in Nordic countries admire them more (Oxtoby, 2009).

In Western countries, lack of professional credibility surfaces at three key points – when social workers interacting with medical and other professionals feel devalued; when social work's porous borders are haemorrhaging as other professions appropriate activities that were once under its exclusive remit; and when only social workers are castigated by the media when practice goes seriously wrong. In some circumstances, the three coincide to the detriment of social work. Social workers' failure to meet expectations in their spheres of activity undermines their position, especially in child protection arenas. Resource shortages contribute much to this state of affairs, but concerns about inter-professional working including the absence of adequate channels of communication between various parties (Laming, 2009) and lack of clarity about the roles of social work in society, are also culpable. All is not doom and gloom because social work has spread globally and those holding the title of 'social worker' can now be found in ninety countries. Social workers in the Global South are more likely to engage in social and community development activities which carry more authority and weight within public consciousness than do the individualized approaches to social problems common in many Western countries.

While the profession is in crisis over its identity and professional status, so are the social and physical environments in which it is embedded. The financial crisis of 2008 has had enormous implications for social work: many countries have cut back on social services provisions and reduced contributions to publicly funded welfare benefits. Because 'Main Street has had to bail out Wall Street', many poor people, and especially those who are more reliant on benefits than paid work such as women caring for children, people with disabilities, and older people – have been particularly disadvantaged. Public expenditure cuts have caused considerable social protest and unrest, especially in countries like Greece and France. Additionally, the new managerialism, including public-sector management initiatives promoted under neoliberalism, have: undermined relational social work (Folghereiter, 2003); reduced practitioners' professional autonomy; and subjected practitioners to managerialist control and a bureaucratization more concerned with protecting agencies from allegations of (mis)using resources than safeguarding the welfare of individuals and communities in need (Dominelli, 2010a; Munro, 2011). Increasing demand for social workers to contribute their expertise in a range of new issues, including disaster interventions and addressing climate change scenarios, is stretching their limited professional resources and capacity to respond to these on a global scale (Dominelli, 2011).

In this chapter, I examine the crises facing the profession and the contexts in which it operates to consider social workers' roles in supporting people through hard times. I explore practitioners' involvement in innovative responses to social problems. Among these, I highlight social workers' engagement in community-based economic initiatives that prioritize the social sphere and physical environment. At the same time, I consider the crises in the social work profession itself as it endeavours to prevent the loss of expertise to other professions, particularly psychology, psychiatry, human geography and criminology. Social work has opportunities to expand into new arenas by becoming engaged in environmental issues and practising green social work within a redistributive framework. The capacity of social work to rise to new occasions is evident in the way in which community workers have energized and organized communities to reject the predations of the market, as has occurred when poor people in the West have been threatened with removal from their homes to make way for a major road; or people on low incomes living in slums like those in Mumbai, India, have been cleared to build prestige housing developments. In such cases, social workers have joined poor people and mobilized action to resist the loss of homes and livelihoods.

A Profession in Turmoil

Professional social work, as distinct from social work as informal caring, was a European invention, created primarily by middle-class women in nineteenth-century Europe (Walton, 1975; Kendall, 2000). In Victorian England, they built on earlier philanthropic traditions fostered by religious institutions, family support networks and volunteers, and aimed to initiate change in individual behaviour alongside improving the administration of charitable support. These endeavours exposed the need for a scientific basis and rationale for such incursions into private family life, and helped establish the realm of 'the social', whereby the private realm became subjected to public surveillance (Lorenz, 1994). This approach became formalized with the establishment of the Charity Organization Society (COS) in the UK in 1868. Shortly afterwards, its favoured model of intervention, the casework approach, became *the* mode for intervening in the lives of poor individuals and families as the newly fledged profession sought to establish its place among other professions. The contradictions between 'care' and 'control' (Parry, Rustin and Satyamurti, 1979) that these initiatives highlight have sat uneasily in social work since its inception and are particularly evident in practice involving counsellors, clinical social workers, probation officers and community workers, as they simultaneously respond to society's demands for well-behaved, law-abiding citizens and challenge its inequalities, structural disadvantages and indifference to the physical environment, while operating within the interstices of a capitalist economic framework that privileges privatization initiatives (Parry, 1989).

The validity of the casework tenets espoused by the COS were questioned by those in the nineteenth-century Settlement Movement that grew out of universities like Oxford which sent students to live and work in disadvantaged communities in the East End of London (Gilchrist and Jeffs, 2001). Its adherents promoted what became known as community development initiatives that focused on structural inequalities, like poverty, unemployment and lack of educational opportunities. Alongside their radical critiques, they sought to educate and mobilize local communities in finding their own solutions to problems, a tradition that Jane Addams spread in Chicago through Hull House and that continues today as empowering and anti-oppressive practice.

The first school of social work to train social workers in a university was opened in Amsterdam in 1899. The one at the London School of Economics began in London in 1901, Birmingham in 1902, and New York and Chicago soon afterwards. The LSE built on the training traditions established by the Women's University Settlement Movement in Blackfriars which was working in conjunction with the Charity Organization Society's School of Sociology that became part of the now London School of Economics (LSE) in 1912. Subjects demanded of those training in casework, youth, group and community work included social administration, economics with an emphasis on applied economics, social history, social philosophy and psychology. Courses in the arts and humanities, e.g., music, featured in the early curriculum. And a period on placement, getting 'hands-on' experience in the field, was also considered necessary. The content of a contemporary curriculum would not seem dissimilar in several respects, although the relative weighting of each course (now called 'module' or 'unit') and the length of time spent on placement might vary. The emphasis on the arts and humanities would currently be found primarily in specific therapies such as music therapy, art therapy, play therapy or reminiscence therapy. Interest in drama would be more likely on community work courses that focus on the dramatic arts to create street theatre or to work with young people.

The young profession had set its sights high, claiming to be able to change individual behaviour and lobby for social reforms that would deal with some of society's most intractable social problems – family breakdown, offending behaviour, disabilities, the difficulties of old age, poverty, unemployment and poor housing. Despite a number of successes, including those of helping people seeking assistance to deal with poor housing, unemployment and other forms of structural inequalities, the profession did not achieve the demise of these social ills (Bolger et al., 1981). Indeed, some increased; others changed their form; and new ones came to the fore. The lack of consistent 'successes' has provided a continuous thread in professional crises since.

From its inception, the profession aspired to be more open and inclusive than the traditional ones of medicine and law. Its ambitions to be recognized as a fully fledged profession were dealt a serious blow when Abraham Flexner (2001), writing his report on the professions in 1915, classified it as a semi-profession at best. This verdict was given on the grounds that social work lacked its own scientific and theoretical bases, and did not have a professional body controlling access to its qualifying programmes. Social work has struggled to overcome the limitations of this legacy since. Consequently, it has been marred by a lack of the professional confidence that features among doctors and lawyers.

Social work was also a dependent profession in that it relied on philanthropists and the state to fund its interventions (Dominelli, 1997). Thus, it lacked an independent professional base from which to affirm its practice and found that it was subordinated to policy imperatives that social workers as practitioners would have had limited, if any, input into and might even disapprove of. This dependency became particularly important in enforcing ideologies that blamed poor people for their predicament, emphasized individual responsibility for structural problems or endorsed racist and white supremacist ideologies, because it became remarkably easy for the state to choose which of the diverse perspectives that existed in social work it would support. So, it was not accidental that the COS provided the model that funders affirmed over the more structurally oriented one expressed by the Settlement Movement. The COS also endorsed the casework or individualizing approaches that spread overseas.

Initially, social work spread globally through the policies of imperial European countries like the UK. In some colonies, for example South Africa, social work was considered the prerogative of white settlers rather than being relevant to the entire population (Sewpaul and Hölscher, 2005). In other locations, particularly among the aboriginal peoples of North America, Australia and Aotearoa/New Zealand, European models of social work intervention were imposed on unwilling populations, usually to their detriment. These often supplanted local indigenous forms of helping (Haig-Brown, 1988). However, local traditions did survive in precarious or underground conditions, and so were never completely wiped out. Even when their disappearance was perceived as a specific danger under interventions aimed at developing social work practice under the auspices of the United Nations (UN) following the Second World War, traditional customs retained their hold. Ioakimides (2011) recounts how this occurred in Greece during this period. Persistent tensions between traditional forms of helping and imposed imperialistic Western models of practice produced countervailing resistance to the latter. This opposition has given birth to the indigenization of social work in many parts of the Global South, and among indigenous peoples that continue to be located in the heartlands of Western countries themselves (Grande, 2004; Gray et al., 2009).

A new concern for the profession is the haemorrhaging at its borders as it loses ground to other professions and becomes incapable of defending territory as its own, particularly in inter-agency working where social workers often feel like the inferior party subject to other more powerful professions such as psychiatry, medicine and law. In the UK, increasing emphases have been placed on inter-agency working in recent decades, but these initiatives have not been without their problems (Warmington et al., 2004). Arblaster et al. (1996) argue that inter-professional tensions arise out of different professional cultures, and lack of equality in the partnerships that form when they work together. Social work, as the less prestigious profession in such encounters, tends to acquire a subordinate role and few social workers dare to question the authority of medical consultants, for example.

Some instances of inter-agency working have strengthened the role of health professionals, psychologists and psychiatrists in social work activities. Of specific relevance in this regard are those linked to disruptive behaviour and mental ill-health. One recent trend in this practice that worked to the detriment of social work is linked to losing Approved Social Worker (ASW) status in the field of mental health work. ASWs were authorized to be one of the two signatories required for the detention of individuals under the 1983 Mental Health Act. This status was protected under the provisions of this Act and only those social workers who received and successfully completed the exclusive training reserved for ASWs could be involved in assessing, supporting and/or detaining those with mental health problems. The ASW was replaced by the Approved Mental Health Professional (AMHP) under the 2007 Mental Health Act. The 2007 Act opened the door for such work to health professionals including psychiatric community nurses, occupational therapists and psychologists, and jeopardizes social workers' dominance in this position (SCIE, 2008).

The New Challenges Facing the Profession

Defining green social work

Intervening in disasters and climate change situations poses new challenges and opportunities for the profession. The onus is on social workers to develop new theories and practices to enhance their capacities in these fields and ensure that these are embedded in both education and practice. I define green social work as:

> a form of holistic professional social work practice that focuses on: the interdependencies among people; the social organization of relationships between people and the flora and fauna in their physical habitats; and the interactions between socio-economic and physical environmental crises and interpersonal behaviours that undermine the well-being of human beings and Planet Earth. It proposes to address these issues by arguing for a profound transformation in how people conceptualize the social basis of their society, their relationships with each other, living things and the inanimate world. It does so by: questioning production and consumption patterns that exploit people and the Earth's largesse; tackling structural inequalities including the unequal distribution of power and resources; eliminating poverty and various 'isms'; promoting global interdependencies, solidarity and egalitarian social relations; utilizing limited natural resources such as land, air, water, energy sources and minerals for the benefit of all rather than the privileged few; and protecting the Earth's flora and fauna. The aim of green social work is to work for the reform of the socio-political and economic forces that have a deleterious impact upon the quality of life of poor and marginalized populations, to secure the policy changes and social transformations necessary for enhancing the well-being of people and the planet today and in the future, and advance the duty to care for others and the right to be cared by others.

In these senses, green social work respects and values the physical or 'natural' environment as an entity in its own right, albeit a socially constructed one, while simultaneously recognizing that people use its bounty to meet their needs. This makes the relationship between people, and their notions of place and space interdependent and symbiotic. Moreover, green social work decries the unequal distribution of goods, services and natural resources and seeks to rectify these. In doing so, green social work, like other progressive forms of social work, becomes explicitly political. Thus, green social work builds on the insights of radical and anti-oppressive social work that identified the political choices people made in organizing social relations in particular ways and the structural basis of inequality in patriarchal capitalist social relationships that have proved incapable of meeting the needs of the majority of the Earth's inhabitants. It differs from ecological or environmental social work, which is a systems-based approach to existing mainstream social work concerns of the person in their environment, usually defined as their social environment, e.g., Gill and Jack (2007). This point is also raised by Van Wormer et al. (2011). Mainstream ecological writings are implicitly political in that they ignore power relations based on existing geo-political social structures, even though these define identity issues, power relations and resource distribution. This shortcoming is amply demonstrated in the works of Germain (1979), Germain and Gitterman (1995) and Pardeck (1996). Some authors raise additional considerations, such as the importance of the spiritual domain in ecological social work. The downplaying of the political nature of environmental concerns in professional practice is evident in the works of Rogge (1994); Rogge and Darkwa (1996), Rogge and Coombs-Orme (2003), Coates (2005), Coates et al. (2006), Borrell et al. (2010) and Rosen and Livne (2011).

In practice, ecological social work initiatives such as those espoused by companies like Kildonan Uniting Care, a company based in Australia, focus primarily on making more efficient use of existing energy resources to enable poor people to get more out of their energy consumption by insulating their homes, submitting grants applications, facilitating the purchase of more efficient appliances, accessing all their benefits and similar measures. Whilst such endeavours are useful, they do not question the structural inequalities that position people on low incomes outside of the market, or its unequal distribution of resources. Nor do these challenge the fundamental bases of an inegalitarian social system currently rooted in a neoliberal capitalist globalization that is primarily concerned with maximizing profits. The ecological social work alliance also remains embedded within ecological social work. However, a group of students at Columbia University in the USA has recently started a Green Social Work Caucus on Facebook and this may produce more innovative approaches to environmental social work that tackle structural inequalities. Also, the Deep Ecology movement organized a conference in Calgary, Canada, in 2009 to encourage social workers and other helping professionals to engage with environmental issues in a more sustainable and appropriate manner. These groups and green social workers can learn from each other by forming strategic alliances.

To work effectively and holistically in the ecological domain, social workers require an extensive array of knowledge and skills that traverse the physical and spiritual domains. Some of these have to be adapted from other disciplines, e.g., the information and data held by physical scientists about landslides, which can be used to avoid the siting of homes in unsuitable locations and thereby prevent deaths in the event of torrential rains or earthquakes that might undermine a fragile ecosystem. The Green Social Work Wheel of Knowledge for Action (figure 1.1) depicts the range of knowledge and skills needed for this work. The material environment in all its complexity depicts an important setting that social workers have to work within and take account of in their practice, during mobilizations of residents, especially at the level of community.

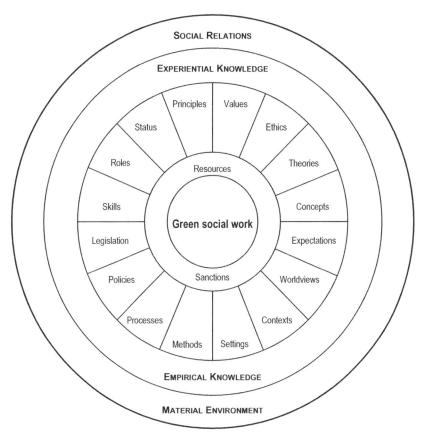

Source: adapted from Dominelli (2002)
Figure 2.1 Green Social Work Wheel of Knowledge for Action

Fiscal, Social and Environmental Crises

Social crises are linked to poverty, low pay, unemployment, inadequate shelter, lack of educational achievement and poor health when individuals, families and communities lack the resources and resilience to promote their own well-being and become reliant on others to provide the necessary resources and/or opportunities to realize their potential. The failure of state responses in this regard highlights the structural nature of these crises and indicates how inequalities caused by society's organization of social relations among its various members, particularly those relating to the production and consumption of goods and services, aggravate these. Structurally caused crises of inequality are not always directly amenable to social work intervention. Rather, addressing these requires social workers to work with others, either by assisting them in mobilizing people and resources to tackle these problems, or by advocating for such initiatives and lobbying policy-makers, corporate owners and opinion-formers to make the structural changes needed for more socially inclusive societies. The complexities encountered can be illustrated by examining offending behaviour.

While I am not arguing that crime is always an outcome of poverty, criminal activity can be both the result of low incomes and the outcome of the lack of resources and legitimate opportunities for fulfilment, especially in the employment arena, when individuals seek to maximize their earnings by other means, to survive in difficult situations. In the absence of other alternatives, young people will become alienated from society and adopt socially unacceptable and ultimately ineffective ways of looking after themselves. While social workers in school-based settings or probation officers can address individual behaviours and induce personal change, they are often unable to address the structural problems of unemployment and failure of educational systems to meet the needs of working-class individuals or people on low incomes, especially those of young adolescent men without prospects or a stake in society.

Case study

Wayne was an eighteen-year-old white working-class man of British origins living in a run-down inner-city estate. He had run away from home at 16 when his father acquired a girlfriend after his mother died, and joined others in a squat. He had a number of convictions for petty theft, driving without insurance, taking drugs, being drunk and disorderly and committing grievous bodily harm. He had spent time in a young offenders' institution.

He told his probation officer he felt alienated and desperate. He saw nothing for him in the future. He had left school at the earliest opportunity, playing truant on numerous occasions until he could legitimately leave. He had no money and, without qualifications, there were no jobs for him. His probation officer thought that Wayne would do well in something he was interested in and asked him what he wanted to do. Wayne said that living in the squat had made him wish that he knew how to build houses because he would like to help people like him live in beautiful homes. The probation officer said he could help him get training in housing construction and carpentry if he were really interested. Wayne agreed and embarked excitedly on his first construction course shortly afterwards.

Wayne's case highlights the importance of environmental rights in the form of decent housing for young working-class men. Once Wayne completed his course, his probation officer intended to work with local employers to obtain a workshop for him to become self-employed. Such action would have solved a structural problem at the individual level and could be used to demonstrate how such an approach – asking Wayne what he wanted and helping him to achieve his goals – could be rolled out to tackle similar issues raised by other excluded groups. But it does *not* address the wider structural problem of the absence of worthwhile jobs at decent rates of pay for young people as a whole. Nor does it address the shortage of decent housing at affordable prices, whether for rent or for purchase.

The link between social ills and poverty is not new (Room, 1995). However, the realization that poverty and the lack of environmental rights in disaster situations are intertwined is a relatively recent insight. In 2005, the consequences of Hurricane Katrina in New Orleans in the USA brought such issues to the fore by exposing the failure of the richest country in the world to care adequately for its own poor people in immediate need of safety, water, food, shelter and medicines; or to respond to the longer-terms needs of victim-survivors in the Ninth Ward for family reunification and long-term housing. Many poor African Americans remain unable to rebuild their housing on plots of land where they had lived previously, many for several generations, because they had lost deeds to their houses in the flooding that ensued, and some families await reunification with missing members, especially children and older people (Pyles, 2007).

I take the view that poverty is a key social disaster in its own right and its presence exacerbates the problems that marginalized groups encounter when other calamities strike, making mitigating risk and dealing with the aftermath of disasters more difficult. Denton (1986) has coined the term '*environmental racism*' to convey this point. Poverty is usually accompanied by the lack of environmental rights, with poor people living in the most degraded environments, in the poorest housing and being disproportionately affected by industrial pollution and natural disasters (Escobar, 1998; Wisner et al., 2011). Poverty also locks poor people in hazardous situations, subjects them to double jeopardy and hinders their capacity to develop resilience or respond adequately. Poverty has particular resonance when emergency planners presuppose a certain level of risk mitigation, e.g., being able to buy household insurance to mitigate risk in conditions of flooding. Because many poor people cannot afford the premiums for such coverage, they face even greater risk when recovering from such events than do their middle-class counterparts (Pyles, 2007).

Replacement housing, whether rented or owner-occupied, becomes problematic for poor people to obtain in the immediate aftermath of a disaster (Comerio, 2002). Their own homes can be difficult to repair or maintain if damaged because such work demands additional resources from households whose budgets are already tightly stretched in covering basic necessities. At other times, landlords become risk-averse and refuse to invest in problematic environments. Thus, victim-survivors often rely on external help to meet their needs and assist in their recovery. Having a roof overhead is vital in coping with the traumas that disasters generate and so rehousing people quickly and efficiently is a crucial priority in disaster relief work. Social workers can monitor this activity and lobby to expedite the fulfilment of people's housing needs.

While 'natural' disasters are rising in both frequency and magnitude, poverty is increasing globally, between and within countries, and disparities in wealth are growing (UNDP, 2010). For example, the number of super-rich individuals or billionaires internationally rose from 793 in 2005 to 1,210 in 2010, despite a serious recession during this period. Meanwhile, the number of people living on less than US$2 per day increased from 2.5 billion to 2.8 billion during the period 2005 to 2007, and most of them lived in sub-Saharan Africa and South Asia (Ravallion et al., 2008). The billionaires are spread over fifty-six countries and hold in excess of US$4.5 trillion between them (Kroll and Fass, 2011). Carlos Slim Helu of Mexico, with US$74 billion, has retained his position as the richest man in the world, having replaced Bill Gates, now the second richest with US$56 billion, several years ago (Kroll and Fass, 2011). Rich women still remain in the minority, numbering 78 in 2005 and rising to 89 in 2010.

A major shift in these figures is the declining position of American billionaires, especially among the men. During this time span, the numbers of Americans in these ranks dropped from one in two, to one in three. Also significant is the growth of billionaires originating from the BRIC countries of Brazil, Russia, India and China. Their figures rose from 130 in 2009 to 332 in 2010. China doubled its numbers to reach 115 billionaires in 2010; Russia has 101, while Moscow has the most billionaires of any city in the world (Kroll and Fass, 2011). Additionally, gaps in wealth are reflected in different consumption patterns between different segments of the population. The wealthiest 20 percentile of people are responsible for 76.6 per cent of private consumption, while the poorest 20 percentile account for 1.5 per cent of it (Shah, 2010). To make visualizing the gap clearer, before the fiscal crisis, Bill Gates had more wealth at his disposal than the poorest 40 per cent of the US population put together (UNDP, 1998). Redressing this imbalance is a priority for those concerned about the impact of consumption on climate change and is an area of social work that could be developed extensively. Social workers attempting to enhance well-being could seek a more equitable distribution of the Earth's material resources within their existing mandate and advocate for its realization.

The current financial crisis has had a deleterious impact on the capacity of poor people to cope with social, fiscal and environmental crises. It has also aggravated existing inequalities as poor people have borne the brunt of the public expenditure cuts and contributed disproportionately to the public funds that bailed out the banking sector. Poverty has also meant that poor people lack the funds to buy increasingly expensive food and fuel. Consequently, hunger and fuel poverty have risen among them while demand for ever dwindling public resources, including subsidized transport, housing, education and health care, has risen dramatically. Rising prices have imposed serious hardship on landless urban dwellers and workers in the rural areas, with disastrous implications for their health and educational status.

The year 2008 saw the sharpest rise in food and fuel prices with oil reaching US$147 a barrel, and poor people rioting about price rises in basic food products in many parts of the world. According to the Office of National Statistics in the UK, food prices rose by 9.3 per cent and petrol products by 25.4 per cent in the year 2010–11. Numbering slightly more than 3 billion globally by 2010, people living on less than US$2 per day usually occupy the most environmentally degraded land in both urban and rural areas. They are unable to grow crops to help them offset the need to buy food, raising serious issues about capacity building in rural areas, especially among poor women (Alston, 2002). Additionally, rural inhabitants may be forced to forgo their future investments by eating grains that were intended as seeds, or selling animals to feed themselves at mealtimes. Huge disparities in wealth and incomes and reductions in publicly funded welfare resources complicate social workers' attempts to address structural inequalities when working to enhance human well-being. However, practitioners can work in conjunction with community groups to tackle these, and join social movements sharing this objective.

While 1 billion people are dying of hunger and a further billion are malnourished, 1 billion over-consume. Additionally, nearly a third (30 per cent) of food grown is not eaten, but wasted. Disparities in food availability can lead to conflict as people migrate away from food-scarce areas to food-rich ones. Food security has become a global issue, especially as current systems of agriculture and distribution of products have proven incapable of solving hunger across the world (Haddad and Godfray, 2011). Food scarcity, these authors argue, means that the technologies represented in genetically modified and cloned food, drought-resistant and salt-tolerant crops are needed alongside organic forms of growing agricultural produce. The technological emphasis of agribusiness is highly controversial. Vandana Shiva (2003) expressed concern that multinational corporations with a vested interest in increasing food production are discouraging the harvesting of seeds from one planting to the next by buying up traditional seeds and food crops to promote the purchase of commercial products. Such practices she calls biopiracy, and argues they have reduced biodiversity in countries such as India.

Others argue that genetically modified foods are bad for the environment and people's health because they compromise natural resilience in crops and the immune system in animals and people. Some suggest that more research is needed before there can be any certainty about the impact of these technologies on people's health and that they should not be used until this has been done and their safety for people and the physical environment is definitively proved (Sinesi and Ulph, 1998; Ewen and Pusztai, 1999). Other opponents of genetically modified food see its production as a moral issue. Haddad and Godfray (2011) write that the moral and ethical questions about the use of such technologies can be ignored. Despite their suggestion, social workers, whose actions are guided by a code of ethics, cannot neglect the ethical considerations that such choices entail. Ethical issues become important not only for their own sake, but also because dissenting voices exist in this field. For example, Devinder Sharma in India has criticized Haddad and Godfray's Report for its limited vision and failure to acknowledge the spare productive capacity that already exists. As he says, 'The world already produces enough food for 11.5 billion people. This [the Report] is just a very clever camouflage for policies which have failed the poor around the world' (cited in Carrington and Vidal, 2011: 15).

Responses to the fiscal crisis: lack of publicly funded welfare services in the age of austerity

The withdrawal of the state from welfare provisions is a worrying development for those accessing services through a welfare state. This is of considerable concern in Western Europe where welfare benefits, including unemployment insurance and pensions, have been attacked by governments seeking to reduce public expenditures, in countries like Greece, Ireland, France, Italy and the UK. These moves are products of shifting economic contexts, namely the fiscal crisis in which governments are looking to civil society organizations and private corporations to plug the gap by encroaching into an arena from which they have previously been excluded, as privatization of welfare services gathers pace across

Europe (Hyde et al., 2003) and in emerging economies like China (Wong, 1994). There are opportunities and dangers in this trend, which has a valued pedigree in North America, although even here, self-help groups funded through the United Way in Canada and the USA are facing substantial drops in the donations they receive. In the UK, the new twist in the privatization agenda is the 'Big Society'. The notion of the 'Big Society' was originally posited by then Tory Leader David Cameron during the 2010 General Election. It was launched on 19 July of that year and was adopted as policy by the Coalition government made up of Conservatives and Liberal Democrats that David Cameron heads. The idea encourages communities to attempt self-empowerment and local decision-making by reducing reliance on government intervention; building on innovative initiatives based on charitable work; entrepreneurial endeavours in both public and social sectors; and responding to public need through grassroots activities (Cameron and Clegg, 2010). It had four showcase local authorities – Liverpool (which withdrew in February 2011), Eden in Cumbria, Sutton in Greater London, and Windsor and Maidenhead in Berkshire.

The Big Society is a vague idea. However, it is grounded in the ideologies of self-help and philanthropic giving which have existed in Britain for centuries. Its formal policy is led by Nick Hurd MP, as Minister for Civil Society, and Lord Wei, as the Prime Minister's Advisor on the Big Society. It is buttressed by the Localism Bill and has a Big Society Bank to fund social enterprises, community groups and voluntary agencies in delivering services. The major UK banks pledged £200 million for these purposes. Some of the Big Society Bank's funding was to be derived from dormant accounts in England. These included those held by poor people, especially pensioners who have put monies aside for a 'rainy day' and do not use them unless such an occasion arises, when they have to struggle through extensive bureaucratic processes to have their money returned (personal communication). Such practices raise concerns about the state's power to raid people's bank accounts without their knowledge or consent. As legitimated unethical behaviour in a democratic society, it raises questions about the state's lack of accountability to individual citizens, especially marginalized and disenfranchised ones. Why did the state not raid the bank accounts of wealthy people to fund community initiatives by passing a law saying that anyone earning over a sufficiently large six-digit figure could contribute a certain percentage to 'good works'? People note discrepancies in the government's treatment of different segments of society, as several residents indicated when commenting on their participation in England's summer riots of 2011 on BBC's *Newsnight* programme, pointing to bankers and MPs who accessed public funds for their own purposes.

The Big Society Network (http://thebigsociety.co.uk/what-is-big-society/) has been created to take advantage of the scope offered by Cameron's policy by bringing together 'silent' taxpayers and increasing their influence on government by:

- empowering individuals and communities;
- encouraging social responsibility; and
- creating an enabling and accountable state.

These objectives were to be achieved through initiatives aimed at increasing localism and devolution; promoting volunteerism; transferring powers to the local level in accordance with the subsidiarity principle; encouraging the formation of co-operatives, charities and social enterprises; and fostering open and transparent government by publishing data on government activities. These sentiments are laudable and to be supported because they encourage 40 per cent of people who believe that they can influence local decisions to do so. Moreover, 1 million community groups and 238,000 social enterprises aim to become more involved in local contexts in pursuit of their key aim of undermining the centralization of state power in the UK. However, as unemployment mounts, there is little evidence to date that there is much substance behind the government's intention to enhance local employment, except rhetorically. By June 2011, unemployment figures reached 2.49 million.

Moreover, the government's Big Society initiatives are insufficient on their own. They do not challenge the existing allocation of power and resources because the initiatives that are engendered by these channel local people's energies within the existing system to smooth out inefficiencies, and simply encourage those at the grassroots level to voice their opinions rather than do more. The failures of large banks to lend to small businesses, be held accountable for their decisions, and contribute their fair share of taxes to the public purse remain untouched by this policy. The government's framing of the Big Society also fails to recognize the variety of voices or opinions that exemplify dissent and diversity in all communities. The government's approach also sets community groups and civil society organizations in competition with each other for scarce resources. In addition, this idea ignores a long and noble history of voluntary work in the UK that has nonetheless failed to deal with the crucial social problems that volunteers have been charged with addressing, because they lack the necessary resources. Resources provided by philanthropic activities and charitable giving cannot meet the substantial need that exists in communities or match funding levels currently provided by the state.

Nor does the government acknowledge the dependency of the voluntary sector on state funding. And, central government has failed to learn from history. The deficiencies of the Big Society idea replicate those prevalent among the philanthropic approaches of Victorian Britain. Key assumptions behind their benevolence included: religious requirements based on the Protestant Ethic; a long tradition of religious charitable works; compassion for other human beings; philanthropists' own personal proclivities and wealth; society's tendency to blame poor individuals for their plight, with particular emphasis on their moral lassitude and unwillingness to work to fulfil their aspirations; and the capacity to ignore the mass nature of the problems encountered by specific individuals. The failure of philanthropy to meet people's needs subsequently led to the creation of the welfare state (Owen, 1982). Why will this approach succeed now? The government has not attempted to answer this question.

The latest twist in philanthropic giving among the super-rich has been initiated by Bill Gates and Warren Buffett in what has become termed 'philanthro-capitalism', or an approach that relies on individual philanthropists pledging to promote 'good works' using wealth that they have accumulated through their corporate activities and interests (Bishop and Green, 2008). These problematic tenets of today's philanthro-capitalists led by Gates and Buffett are echoed in Cameron's credo for the Big Society. Yet, while Victorian philanthropists focused primarily on their own back yard; the main theatre of operation of today's billionaires tends to be in the Global South rather than in their own countries, making this a contemporary version of *noblesse oblige*. Together, two of the richest men in the world, Gates and Buffett, have convinced forty fellow billionaires to pledge US$238 billion in charitable donations, though whether this sum has been delivered in its totality is less clear.

I am unable to define this move as anything other than a continuation of a well-established Western tradition of turning individual philanthropy into corporate philanthropy, with its attendant tax relief, or public subsidies to private bodies by another name. The Gates Foundation currently has US$33.5 billion to disburse on HIV/AIDS and to promote a sustainable green revolution in agriculture. Both priorities might produce considerable profits for this organization as it proceeds in fulfilling its mission. It is anticipated that in ten years the Gates Foundation will have a GDP bigger than 70 per cent of the world's nations (Bishop and Green, 2008). This concentration of wealth in the hands of a few unaccountable individuals seems unethical. There is a moral argument around why super-rich people should be allowed to decide the priorities to be followed in other countries when they do not pay taxes proportionate to their wealth either in these places or in their own countries of origins, and their enterprises do not pay workers wages that allow for a substantial rise in their standard of living.

These weaknesses in philanthropic giving led the Fabians in Victorian London to favour the creation of a welfare state funded through social insurance, under conditions in which workers were entitled to receive benefits as a right (Webb, 1909). This message was strongly endorsed by the then workers' movement, mainly the trade union movement and subsequently the Labour Party, the organization that workers formed to promote their interests in government (Webb, 1918). These gains and their associated notion of rights to benefits and services that promote people's welfare or support them at times of need, on the basis of solidarity and a pooling of risks, are being undermined by the thrust towards the Big Society, the current public expenditure cuts and the impact of the loss of state funding for public goods, despite Cameron's avowals that Big Society activism will address people's needs.

Other sceptical voices about the potential of the Big Society to realize its objectives were raised by Labour Leader Ed Miliband who dubbed it 'a cloak for the small state'; Anna Coote of the New Economics Foundation predicted that it would produce a 'diminished society, not a bigger one' and that Cameron's big idea was about 'privatising the welfare state on a massive scale'. Meanwhile, trade unionists like Brendon Barber echoed the view articulated by Mary Riddell that it harboured 'the sink or swim society', and suggested that the Big Society revealed that the Prime Minister's 'ideal society was Somalia'. UNISON's David Prentice argued that volunteerism provided a 'cut-price alternative' to decent public services. And, Steve Bell claimed in the *Guardian Weekly* that the motto for the Big Society would be 'From each according to their vulnerability, to each according to their greed'.

These criticisms notwithstanding, several voices responded positively to Cameron's ideas. For example, Ben Rogers in the *Financial Times* reported that social ills could be solved if 'residents and citizens played their part'. Ben Brogan at the *Daily Telegraph* thought this allowed 'society [to be] rebuilt from the ground up'. Ed West at the *Daily Telegraph* was supportive, although concerned that the 'socialist ideology' of the British public would prevent the Big Society from taking off. On the upside, it was hoped the Big Society might promote initiatives that would increase alliances between the working and middle classes, the latter of whom are much better at mobilizing to defend their specific local interests than the former (Mott, 2004). Class inequalities can be reinforced through such relationships unless middle-class organizers desist from exercising top-down power relations within working-class communities (Rose, 2000). Particularly important in this regard is avoiding the 'false equality trap' whereby middle-class people speak from a position of knowing, arrogance or superiority, or all three, and belittle the contributions made by local working-class residents, especially when community-based suggestions might differ from theirs (Barker, 1986; Dominelli, 2006). Knowing members of the community and its resources (or lack thereof), developing relationships of trust and achieving early, if small, successes in meeting locally expressed goals are important in developing local collective, community action.

The savage cuts in public expenditures carry considerable implications for the viability of Big Society endeavours in contemporary Britain, relying as they do on voluntary involvement in the community. Local authority budget cuts for 2011/12 will hit the voluntary sector hard and result in an average reduction of 19 per cent in overall funds disbursed to voluntary agencies. Leeds, Liverpool, Manchester and Sheffield alone have cut their contributions to this sector by £43.9 million. A few councils are bucking the trend. Newcastle City Council (NCC) is one of them. It has added £3.5 million to its voluntary-sector budget to ensure that its workforce of 7,000 remains in business and generates £70 million in revenues. But NCC is making £45 million in cuts elsewhere (Plummer, 2011).

The government's refusal to empower communities through substantial and continued injection of public funds that focused on collective action makes the Big Society a highly conservative concept that is likely to reproduce existing resource inequalities and hinder improvements in the quality of life in poor communities. Thus, I define Cameron's Big Society as an idea that demands that people become self-sufficient in meeting their own needs, primarily in the form of *delivering their own services* within the context of reduced publicly funded welfare resources, by producing innovative responses to current and projected social problems within existing resource constraints and political parameters. I conceptualize it in this way despite Cameron's desire to see the devolution of central power because transformative social change cannot occur at either local or global levels without a significant transfer of resources, decision-making powers and authority capable of tackling structural inequalities; holding accountable the financial sector and super-rich elites for their failure to pay their share of taxes; working for the good of the whole of society, not only the privileged few or large shareholders; and restructuring the economic system so that it serves people rather than the other way round. The Big Society concept also takes the physical environment for granted. Care of the environment will transpire only if local people prioritize it and embed this concern in all their activities.

The conservative nature of the Big Society is further evidenced in injunctions Eric Pickles (2011) as Communities Secretary issued to local authorities in April 2011. These were to:

- avoid disproportionate cuts in funding;
- provide three months' notice if cutting funding or terminating it;
- work with organizations to determine the shape of future services; and
- engage organizations in discussions on how to improve service delivery or develop innovative alternatives to existing provisions.

Such rulings are likely to become empty rhetorical gestures given the tight budgetary constraints on local government. But they can be used to shift the blame for the UK's economic woes away from central government and the financial sector in the City of London onto local communities.

The evolution of the Big Society may prove instructive, with lessons about what a state can abstain from doing, resonating and being picked up elsewhere. The lack of equity for marginalized groups, disregard for social and environmental justice, and focus on individual and community failings, while doing nothing about structural failures, the inability of current financial institutions and resource allocation and distribution systems to provide jobs, social and material security, and protect the physical environment, are dangerous for people and Planet Earth. Cameron could take a leaf from Bissio's (2011) Report on how social policy can be reformulated to address issues of social and environmental justice. Yet, the idea of organizing at community level without government interference is a good one. Social workers or community workers can use it to develop alternative local institutions. Practitioners should ensure that pre-existing agendas based on exploiting people and the natural environment are not inappropriately driving local action and that there is a real basis for forming alliances across issues that might otherwise divide community groups (Warren, 2001). Social workers assessing the implications of these responses when engaging local residents in providing services for their communities would juggle these conflicting realities to cater for people's needs and develop transformative responses in the absence of publicly funded facilities. Reconciling diverse interests and operating within a social and environmental justice framework is within their professional remit (McKinnon, 2008).

Urbanization and Slum Clearances

Urbanization has placed pressure on land, a key resource for the Earth's inhabitants, flora and fauna and built infrastructures, including housing, transportation systems and communications. Land, as a scarce resource, becomes very valuable as demand for it rises. Land scarcity pits the needs of different groups of people against each other, especially as land values in city centres rise as development proceeds to meet housing, transportation, communication and other needs, and to make a profit for developers and others. For example, some Roma people had lived in downtown Naples for twenty years. They became engulfed in a maelstrom when city developers linked to the Mafia decided that they wanted this land for their own purposes. Mafia supporters set about accusing the Roma of leading lives of crime and putting at risk the lives of local Italians. A bitter conflict broke out and the Roma were forcibly evicted to make way for the construction of luxury housing (OCSE, 2009). The slum dwellers of Mumbai faced similar difficulties when their homes were bulldozed to make way for luxury apartments. In one situation, even the home of Robina Ali, a child star in *Slumdog Millionaire*, was destroyed.

Poor people who may have historical rights to plots of land that become valuable commodities in such scenarios may find that they lose out in struggles over land. More powerful groups may seek to evict them, often using force. This occurred historically during the Land Enclosures in England and Scotland, when peasants and small tenant farmers were evicted to create large mechanized agricultural estates that enabled the owners to make profits from the land. Those working the land became destitute. Some emigrated overseas to colonies like Canada and Australia. Those who remained in the UK went to the cities, seeking work that was often unavailable, and became part of a landless proletariat (Thompson, 1963), often being unable to access poor relief that was linked to residence in a particular parish and subsisting to the best of their ability with the limited resources at their disposal. Others had recourse to workhouses where an impoverished existence was provided to ensure that paupers, as they were called, were compelled to seek work, even in conditions in which employment opportunities were scarce and wages a pittance. The lives of hardship that featured in the absence of the welfare state are being replicated in the megacities of the Global South today where welfare states are also absent.

Without the welfare state, the fate of dispossessed people rested mainly in the hands of charities, religious institutions and family members, who were expected to care for each other as best they could. This remains the pattern in the Global South, except that secular non-governmental organizations (NGOs) and other civil society organizations (CSOs) have been added to the mix. In the Global South, the impact of NGOs and CSOs with overseas origins has proven problematic in several respects. Often these bring customs and traditions that are at variance with local ones and may enforce priorities and working practices that are inappropriate and resented by recipients (Mohanty, 2003). Additionally, overseas NGOs only provide jobs for some local people. They are usually paid higher wages than those prevailing in the local labour market and distort its functioning. Finally, their remit for intervention may be short-term and this creates problems of sustainability when overseas NGOs depart (Shivii, 2006). This creates ambivalences, tensions and contradictions that local residents and overseas workers have to address if they are not to perpetuate globalizing practices that are exploitative and reinforce neo-colonial forms of social relations.

The complications of urbanization and competing demands for land can exacerbate the problems that social workers encounter in practice, as the case study below indicates.

Case study

Sandra was a community development worker committed to developing sustainable communities and using the insights of green social work to support a community group that was opposing the construction of a four-lane highway through a close-knit rural village because it divided the village into two. Local people felt they did not want this to happen because they had family and friends who would be divided from each other and who would not be able simply to cross the road to offer help when needed. The company that was building the road refused to listen to the community's concerns, although it offered to build a pedestrian bridge across the road as the solution. The residents rejected this option because they felt it: destroyed their views by 'urbanizing' their landscape; required them to invest more time and energy getting across the road than previously; and imposed additional expenses on those who were frail or had mobility problems, e.g., purchasing scooters to help them cross the road. Using motorized vehicles became possible when the company said that it would substitute access by ramp rather than steps, when the mobility problems were pointed out. However, this did not address the issue of affordability.

When the community discussed its options for breaking the impasse reached with the company, they decided they would use a holistic approach to cover all the actual problems and costs that the proposed roadway would inflict on their village. They would not benefit from its construction, as access for local residents was 10 miles away. They would have to put up with dirt and noise while it was being built. There were changes that would ensue in perpetuity, like the inconvenience of getting from one side of the village to the other over the footbridge and loss of landscape amenities, including ponds that harboured protected species.

The community group enlisted Sandra's help in devising an empirical study that would enable them to articulate all costs precisely and comprehensively. To undertake a holistic assessment that went beyond a usual cost–benefit analysis and to consider quality of life issues, she connected them to artists, economists, mathematical modellers and other professionals. Their tasks included determining how much time they would spend getting house to house crossing the road over the pedestrian bridge as opposed to the time spent currently without it. The group also ensured that differentiated needs were taken into account, e.g., those who had frequent visits to sick relatives, friends and frail elders, and women with children going to the local playground, as they would have to invest more time getting across and this would add to their caring burdens. An artist's drawings were helpful in assessing the visual impact of bridge and highway on the landscape.

The tenor of the community group's meeting with the company changed once they exchanged views backed up with empirical data to endorse them. In the end, the village was by-passed with minimum damage to the environment, to the satisfaction of both parties to the dispute.

This case demonstrates how community social workers drawing on green perspectives can empower communities in power struggles with more powerful others if they have access to the skills and expertise needed to make their case and insist that they also have rights, including that of being heard.

Conclusions

Social work suffers from a professional crisis of confidence that means that it is not proactive in either defining its borders or developing in new directions. However, individual practitioners are constantly innovating to address the needs of those with whom they work and embedding their activities in new theories and approaches to practice. Green social work provides opportunities for innovation that can deal with issues of poverty, urbanization and a holistic approach to sustainable development.

Green social work provides insights that enable community groups to empower themselves by drawing on professional expertise from a range of disciplines that have the knowledge to develop holistic empirical appraisals of their predicament. The case study considered above highlights how community development work can move in new directions when challenging decisions made by powerful corporations and helping communities document their case so that it can withstand scrutiny. This means drawing the expertise of other professionals into the practice circle. It also empowers communities by enabling them to develop their own solutions to problems and recognizing that they have the right to realize their rights as citizens. Sustainable community development practice is good practice that includes protecting the environment by caring for it when meeting human needs.

3

Reclaiming Industrialization and Urbanization for People

Introduction

Industrialization has produced pressures for centralization and the development of urban environments and built-infrastructures that are driven by the creation of opportunities for business to make profits and people following the employment prospects that these present. Development based on urbanizing the environment has created population flows from rural areas into urban centres, and intensified pressures on the capacity of the physical urban environment and its resources to meet people's physical, social, recreational, cultural, political and economic needs. The United Nations has predicted that the trend to urbanization will intensify to create 33 megacities mainly in the Global South by 2015. The majority of them will be in already fragile coastal areas (Nicholls et al., 2007).

The numbers involved in megacities and the area that each city encompasses are contested. A megacity has been defined by the United Nations as a city having more than 8 million inhabitants. This figure is now dwarfed by the size of some, e.g., Mexico City has a population of over three times that number. Recent discussions place these cities at 10 million people each. Whatever figures are considered, urban planners agree that these cities are complex and problematic as well as dynamic and exciting. Their growth and constant renewal are fed by the migration of rural peoples, those moving from other cities and those coming from overseas. Rural–urban migration today is widespread in China, which currently hosts more megacities than either the USA or Russia. In China, those who migrate to cities without official permission are denied settlement rights in them, and so live in extremely precarious conditions.

Except for inland cities like São Paolo, Mexico City, Delhi and Beijing, the location of megacities in coastal areas increases their vulnerability to flooding, and other natural hazards such as hurricanes, that might be exacerbated by climate change (Nicholls et al., 2007). The number of vulnerable megacities is likely to rise further as populations in Africa and Asia grow and a number of today's smaller cities become the megacities of the future. Those living in such cities will also have to deal with the problems of poverty, as most megacities contain heavily populated slums with: poor access to communication systems, water, utilities and sanitation infrastructures; inferior-quality housing; places where infectious diseases can spread easily; deteriorating air quality; and various forms of environmental degradation produced as a result of increased urbanization and growing population numbers in limited spaces.

The urbanization approach has had enormous implications for poor people's well-being, often without their having a say in the development of initiatives that yield such outcomes. Yet, they often bear the direct costs in poor living conditions and a degraded physical environment. Whether this emanated historically from the Enclosures that left rural people in the British Isles landless or destitute in the big cities of Victorian England like London, or in the current period for the villagers in the shanty-towns of the contemporary megacities in the Global South, the end result has been the same: a drop in the quality of life for once self-sufficient people in the short term, and a dependency on their ability to sell their labour in the long term. At the same time, social relations became more hierarchical and differentiated, both within the home where men came to dominate and in the workplace where employers set the terms to be obeyed by those fortunate enough to hold paid employment. In this system of social organization, largely described by the term 'capitalism', poor communities have had difficulty holding those with wealth accountable for their behaviour and the decisions they make, even though these carry negative implications for poor people's lives and the communities that they inhabit. Lack of accountability regarding the decisions that the owners and managers of global multinational corporations take remains a problem in the system's latest expression, namely neoliberalism. Within the context of economic globalization, the unaccountable owners of contemporary corporations may be located far from the areas most affected by their decisions.

In this chapter, I consider how social workers can work with people at the local level to develop more sustainable and life-enhancing forms of urban living that require accountability for the actions taken by all the stakeholders involved in a local area. At the same time, I acknowledge how local schemes can have an impact at national and international levels and provide important insights for improving situations through collective action. Examples drawn from both the Global North and the Global South explore issues like the formation of locally accountable financial institutions such as micro-credit ventures, credit unions and local area networks, including the creation of local exchange schemes to avoid monetary transactions. And I use case studies that include optimistic scenarios for the development of a sustainable, interdependent world.

Urbanization, a Defining Feature of Industrial Capitalism

An abundance of cities is a fairly recent global phenomenon. The United Nations estimates that, in 1800, only 3 per cent of the world's population lived in cities. The 1990 Census in the USA declared that more than 90 per cent of its population lived in cities of over 1 million inhabitants. Moreover, the country hosts two of the largest city conglomerations in the world – New York and Los Angeles. Western societies have gone from being predominantly agrarian, pre-industrial societies to urbanized industrial ones in about 150 years. In the UK, 43 per cent of the population was urbanized at the beginning of the industrial revolution, compared to 65 per cent today. The UN report *World Urbanization Prospects 2009* claims that urbanization will have risen from 13 per cent, or 220 million, of the Earth's inhabitants in 1900 to 60 per cent, or 4.9 billion, by 2030 and predicts that urbanisation will proceed at an even greater pace after that. Urbanization, population growth and technological developments, including those in agriculture, came together to produce contemporary cities as we know them. Urbanization occurred as rural farmers, unable to make a living in their home environments, sought jobs and enhanced living conditions in cities, a trend that pertains in the current historical conjuncture. This has been added to by the growth of cultural, social, recreational, political and economic opportunities in the cities that attract people to their bright lights, and the economies of scale in building infrastructures to serve large numbers of people. But there is a price to be paid when concrete replaces green fields, and this complicates city living. It also problematizes how the price for such developments should be calculated, both for opportunities gained and opportunities lost, in the trade-off between the needs of people and those of the natural environment.

At the same time, cities do not offer all people a smooth transition from rural to urban settings. Housing, employment and other provisions may be lacking or too expensive for rural dwellers to access. The mass migration of rural labourers into cities that lack infrastructures such as adequate sanitation, lodgings, transportation and jobs to support their number can exacerbate their difficulties. Limited or non-existent schools and health facilities can be added to this list. Eventually the absence of such facilities creates slums that are characterized by overcrowding; unemployment; the spread of infectious disease like tuberculosis, diphtheria, cholera, typhoid and bilharzia; and crime. Vivid portrayals of the social problems historically faced by working-class residents in British cities rose from the pens of Gareth Stedman Jones in *Outcast London*; Friedrich Engels' *The Origins of the Family, Private Property and the State*; and Thomas Chalmers' concern to improve such conditions in Scotland, in the *Godly Commonwealth*. The advent of public sanitation, clean drinking water piped into homes in the cities and the construction of (social) council housing did more than anything else to alleviate the problems of urbanization for people living in poverty and squalor. Whilst poverty, slum dwellings and disease were common features in Victorian cities like London, Birmingham, Manchester, Newcastle and Glasgow, they are not unknown in the burgeoning megacities of the Global South today, especially those like Mexico City, Rio de Janeiro, Jakarta, Mumbai and Dhaka.

Authorities have attempted to control rural–urban migration through various policies. Industrializing Britain sought to limit the demands of destitute mobile populations for public assistance through the 1601 Elizabethan 'Poor Law', which gave limited relief to those who fulfilled stipulated criteria for assistance. Its restrictions included residential clauses, e.g., assistance was to be claimed from the parish in which a person was born. This arrangement was modified somewhat in 1875 through the Speenhamland system, which authorized the provision of outdoor relief for poor rural people, to cushion them from rising food prices. The provisions of Speenhamland became a means-tested wage supplement linked to changes in bread prices, a food staple at the time. The conditions for receipt were strict and the land-owners who were expected to fund the scheme ensured that poor claimants sought work or went into the workhouse rather than claim public relief. Success in diverting claimants in this manner meant that the cost that the landlords bore to maintain the scheme was minimized. Arguably, the policy also helped to keep wages down by subsidizing them through disbursements that attempted to curb the impact of rising prices for bread on poor folk.

From this point, the Speenhamland system was in place in some shape or form, as an adaptation to the Poor Law Reform Act of 1834, for nearly forty years. This approach has echoes in the economies of rising powers like China and in other transitional countries aiming to curb the public expenditures available to support poor people. A main concern is that such policies subsidize low wages.

The growth of megacities in the Global South has intensified urbanization and centralization. Although the existing infrastructure and services are unable to cope with the numbers of people involved, people continue to come. Mexico City, one of the largest cities in the world, faces daily increases to its figure of over 30 million inhabitants. In China, megacities are becoming prominent and growing in size. The government is deliberately constructing some of these, e.g., Tianjin, on the grounds of benefitting from the economies of scale in their laudable struggle to eliminate poverty in that country and to meet its obligations under the Millennium Development Goals (MDGs). Although Schumpeter (1935) raised concerns about the costs of sustainable economic change and Schumacher (1974) questioned the appropriateness of centralized urbanization several decades ago, the latter's idea that 'small is beautiful' did not catch on among policy-makers. Thus, their interest in ensuring that the costs of urbanization did not undermine the need to preserve green spaces for people, plants, animals and the physical environment on Planet Earth did not become as popular as anticipated.

However, it is not only government that has an interest in and responsibility for urbanization. Employers do too. The combined failure of government and employers to provide the resources necessary to cater for the needs of large urbanized populations has resulted in huge slum areas or *favelas*. These then become sites typified by extreme social problems, including poverty, prostitution, substance misuse and gang violence, sitting alongside unemployment, disease, street children scrabbling for a living, poor educational prospects and other forms of social exclusion that have become synonymous with the term 'megacity' in the Global South.

Industrialized countries, made up mainly of those in the West, also have their megacities – including Britain's London, arguably the archetype of such urbanization; New York in the USA; and Tokyo in Japan. That poor people in the West live in relatively better conditions than those who are destitute in the Global South does not mean that these cities have succeeded in meeting the needs of all members of their populations, as required by their governments' obligations under the Universal Declaration of Human Rights, in which Articles 22 to 27 support people's right to food, clothing, shelter, education, health services and income support in times of need. Social workers are expected to deal with the problems encountered by poor urban dwellers, intractable as these are, and to do so without adequate employment, educational, housing, health and other resources at their disposal.

Decaying infrastructures, especially in transportation and sewage systems, are also causing problems, especially as private companies seem to be investment-averse. In the UK, the privatization of public utilities including water, electricity and gas did not yield the promised returns on infrastructural investments or lower tariffs for consumers (Newbery, 1997; TW, 2011). So, infrastructures developed in the Victorian era creak along, trying to meet the heavier demands of modern consumer lifestyles. In Montreal, Canada, the lack of investment in the arterial highways around the city resulted in bridges and roads collapsing in the summer of 2011, much to the dismay of its residents who are now calling for an injection of federal government cash to correct the matter. Levesque and Mathieu (2011) argue the issue is one that affects all Canadian cities, where 80 per cent of its current population lives. Brox (2008) estimated that renewing Montreal's transport infrastructure would cost CDN$200 billion. Building the necessary new projects would consume CDN$72 billion and repairing existing stock would use CDN$123 billion of this sum. He also suggested that not attending to this would cause Canada to lose its competitive edge as a trading nation because productivity would decline and production costs would rise. His arguments are all about economic priorities, not social needs, which get barely a mention in such discourses. Can economies that favour people and the natural environment make better use of the vast sums being invested to cater for sectional interests? It is a question that social workers and policy-makers should be asking.

Competition for jobs, housing and other resources in cities may be intense, and often the interests of one group may be pitted against those of another. For example, developers evict slum-dwellers to invest in prime real estate in city centres, as has occurred in places as diverse as Mumbai and Naples. In 2008, the Roma in Ponticelli, Naples, were violently evicted from a city-centre community that they had occupied for several decades because the city's ruling elite wanted to construct buildings for its own use on that land and the Roma people were in the way (OCSE, 2009).

In Mumbai, even the global fame of Robina Ali, a young star in *Slumdog Millionaire*, did not prevent her home from being cleared to make way for luxury housing. Land prices in cities like Mumbai are now higher than those in Manhattan (Jeong, 2011). In Mumbai, the Dhobi Ghat site, where mainly men from the Dhobi caste run a laundry business, is being threatened by developers. The site, though small, is located in Mumbai's financial district and allegedly worth US$200 million. It has been owned by the Dhobis since 1958 when they secured a collective agreement with the city granting them ownership. At the same time, migrants from all over India are joining the Dhobis, placing more pressure on the site's poorly maintained built infrastructures, particularly those of water and power, and exacerbating the lack of housing for people who are earning US$10 to US$20 a day washing, drying and ironing clothes (Jeong, 2011).

Evictions usually involve rich people making decisions that have a deleterious impact upon the lives of poor people who often resist such moves without achieving their goal of remaining in areas that had previously provided their homes. Place and space have become contested terrains in industrialized and industrializing societies because, along with providing sites for making profits, they are closely linked to people's sense of continuity with the past, individual and collective identities and belonging that are affirmed by occupying a terrain with ancestral links and where others approve of them and they approve of others. Belonging and continuity provide feelings of power and entitlement because local residents define these terms to enable them to have a say in who is accepted or rejected in their community. This reality in local space contrasts sharply with feeling disempowered and marginalized within wider society. However, current power politics turn landless people into an internally displaced population with few options for settlement elsewhere. Their internal displacement can be within the confines of the city itself, as in the case of Mumbai, or in the region, as indicated by the Roma in Europe who have been shipped out of countries such as France and Italy to their place of origin in Romania (ERRC, 2007), or displaced from their homes of ten years with nowhere to go as occurred in Dale Farm in the UK (Townsend, 2011). In Italy, government officials and the media fed public hysteria against Roma peoples, despite European and Italian legislation on equal rights (ERRC, 2007). Such areas are often exploited by organized crime syndicates who prey on slum-dwellers for trafficking people and organ removal, again to the advantage of those who have money to pay for them. Most people living in a shanty town called Baseco, near the dockside in Manila in the Philippines, have sold one of their kidneys for cash, receiving as little as US$481 per kidney (Padilla, 2008). People should not be reduced to these levels of desperation but the existence of such predatory practices indicates generalized indifference to their plight, which has been well known for years. Organ-selling scenarios, sometimes using violence to coerce 'donors', were vividly displayed in *Slumdog Millionaire* as it narrated the story of how poverty provides the terrain that unscrupulous individuals and criminal gangs exploit for personal gain. In a global world, poor residents are drawn into the process of enriching themselves through the sale of organs and trafficking of human beings (Nagle, 2008). In these narratives, the body becomes inscribed as a profit-making site for these criminals and it becomes reconfigured as a *place* from which the privileged few extract profit.

Cities destroy natural habitats by utilizing land for buildings, streets and other purposes. This can have deleterious consequences for biodiversity and the natural environment, although several resilient species have adapted fairly well to these new conditions, e.g., foxes and rabbits in the UK; deer, bears and coyotes in North America. Yet, there is severe competition between the needs of human beings and other living things for the limited overall space available in cities. The replacement of living ground cover with concrete can increase the likelihood of flooding as the land underneath cement can no longer provide natural drainage sinks and built infrastructures prove incapable of dealing with excess water during extreme wet weather events, including torrential monsoons. In Mumbai, for example, flooding in 2005 caused US$2 billion in damages. These and more recent floods have overwhelmed the capacity of the city's drainage system, including the river Mithi and its sewers (Hallegatte et al., 2010).

Urbanization in the form of slum dwellings along the banks of the Mithi, the disposal of both domestic and industrial rubbish in its waters and natural silting mean that a massive clean-up effort has to be initiated to enhance the river's capacity, to reduce future flooding hazards and eliminate the stench emanating from it. Social workers can engage with local communities to find alternative ways of disposing of waste, and alongside community groups, advocate among government and private entrepreneurs for the Mithi river to be cleaned up. To gain support for this, they could build on the recommendations of an OECD report by Hallegatte et al. (2010). Similar issues arise elsewhere. In the UK, unprecedented rainfall led to the 2007 floods in major parts of the country. According to the Pitt Report (Pitt, 2007), the limits of existing flood alleviation schemes were exceeded and inadequate drainage systems in urban settings compounded matters.

Social workers in the UK were involved in responding to people's needs in evacuation shelters, finding temporary accommodation, and addressing issues raised by the longer-term impact of these events. Building resilience in critical infrastructures, including roads, is the responsibility of civil servants located in the Cabinet Office. Practitioners can help community groups monitor progress and assist in initiating complaints procedures if work does not proceed according to the Pitt Recommendations for advanced planning and early warning systems to be put in place to mitigate damage to life and property. Social workers can also help people to identify areas that the insurance industry has declared uninsurable and therefore, not recommended for building upon, e.g., flood plains. Such decisions carry implications for housing construction, determining locations where people can reside safely and protecting fragile ecosystems. Social workers should also become familiar with the Flood Rescue National Asset Register in order to assist in matching need with available resources in the event of a flood. They can also develop good links with officers in recently formed National Hazards Teams (NHTs) to ensure that the best preventative measures are developed for each particular area. Social workers also have important roles to play as part of the emergency response services trying to minimize disruption in specific hotspots.

NHTs have already identified those areas that could suffer from failures in communications, emergency services, energy, finance, food, governance, health, transport and water, during times of disaster. Social workers have a responsibility to find out the location of these areas and engage local people in prevention and mitigation activities to enhance their resilience and capacity to survive if a calamitous event does occur. They can also assist in fleshing out the Sector Resilience Plans through which the government intends to reduce vulnerability to catastrophic 'natural' hazards. Government teams and utilities companies tend to focus on the big picture, e.g., flood-proofing the walls at the Mythe Water Treatment Works in Gloucestershire that flooded in 2007. Social workers can play a key role at the micro-level, e.g., supporting residents in registering online for the flood warning services, and coordinating the activities of different players to ensure that robust plans and disaster response kits, such as flood kits, are devised by communities. This could involve them in linking knowledge about the condition of the local infrastructure held by different agencies, such as the Environment Agency and local authorities and physical scientists, to build upon traditional knowledge and daily life experiences held by local residents. Social work students, and particularly those interested in community development, could become involved in this work through either a placement or a piece of research linked to their dissertations. They could also join students from other disciplines, e.g., geography and engineering, because they have different, but relevant, information to bring to the disaster mitigation table. On the flood front, these students could contribute to the Pitt Review Monitoring Task Groups, Scrutiny Committees and other community fora, such as the Local Resilience Forum (LRF) set up by local authorities. The LRF liaises with the Regional Resilience Team (RRT). There are nine RRTs for forty-three LRFs in England. Some authorities, e.g., Hampshire, have established an adverse weather multi-agency response team located in the Hampshire Police control room. The police and the firefighters are among the first members of the emergency response team to arrive on the scene during a disaster. According to the 2004 Civil Contingencies Act, local authorities are Category 1 responders with major responsibility for meeting the needs of those affected. Social workers usually come in later when evacuation has safely taken place. But they can assist in pre-disaster planning and the development of preventative measures that can mitigate risk before and during a disaster, as well as contribute to rebuilding communities afterwards.

Additionally, the countryside usually bears the price of meeting the needs of city-dwellers. In the UK, this has resulted in the loss of hedgerows and bird habitats on farmland to increase acreage for growing food for urban populations. Agricultural land decreased from nearly 20 million hectares to about 17 million between 1961 and 2005 (Rounsevell and Reay, 2009). About 76 per cent of land in the UK is used for agricultural production, much of it under the control of agribusinesses. Since 1990, about 15,000 hectares per year have been developed for non-agricultural purposes. And there is a constant trade-off being made between population movements, including growth in the size of populations, and the need for land to grow food, build housing and move commuters from their homes to places of work. To reduce pressure on green-field sites, the British government is exploring the use of brown-field sites for housing and anticipates that 75 per cent of the 4.4 million homes needed by 2016 will be built there.

The proportion of the British population involved in commuting is the highest in Europe, and thought to be about 25 million. Commuters consume 4.3 million hours per day getting to and from work. The distance of the commute rose by 17 per cent between 1993 and 2003, and the time spent commuting rose yearly until 2006 when it began to decline. The Trade Union Congress (TUC) reckons this reduction is linked to home working, and suggests that it costs about £339 million in working time to commute (Khan, 2008). Social workers can raise questions about the appropriateness of this way of organizing work and its costs to the environment and people's health. Transport plans are predicated on responding to the needs of large urbanized and highly centralized cities and they increase the burdens being carried by the bio- and eco-systems (Khan, 2008). This burden is additional to the costs borne by people in terms of less time being spent with families or resting, as these journey times eat into these spaces and transform this daily routine into unpaid work time as people continue working on long-distance commutes. In the UK, long-distance commutes are not uncommon. The difference between time spent performing tasks linked to paid employment and time spent enjoying family relationships has been termed the work–life balance, except that for most people it is an imbalance that favours the interests of employers. Moreover, the locations that people commute from have become little more than dormitory centres and so community relationships suffer as a consequence of work obligations driving other social relations.

Commuting as a way of life is an energy-intensive mode of existence and integral to centralized economic activities linked to urbanized centres with industrial structures and services. Decentralization has proved difficult. In the UK, successive governments have promised to relocate key public-sector agencies outside London and provide incentives for private industry to move out of the capital too. However, except for the occasional genuflection in this direction, most economic and employment opportunities remain in London. The spatial dimensions of urbanization show the complex networks and linkages, ranging from communication to transportation within the built infrastructure to those involving social relationships and people's interaction with each other and the flora and fauna that surround them. This complexity has led many to question the appropriateness of urbanization as the solution to the imperative of taking all people out of poverty and raising the standards of living for those who have been excluded and marginalized. From this vantage point, urbanization, based on the capitalist model of industrialization, is unsustainable (Brahic, 2007). Given that the industrial revolution began in the eighteenth century in a small island nation that had 6 million people, many of whom were excluded from its benefits despite providing the labour power that drove it (Thompson, 1963), the model's limitations and inability to meet the needs of the billions that now inhabit this planet should not come as a surprise.

Modern urban living also consumes enormous amounts of water for domestic purposes and for appliances such as washing machines and dishwashers to perform household tasks; electricity becomes increasingly used for gadgets often left on 'stand-by'; fossil fuel is utilized to transport people to workplaces; green spaces are losing out to concrete for high-density housing, services and parking lots. Domestic energy needs account for about 40 per cent of current greenhouse gas emissions (Löscher, 2009). The transportation of goods, services and people is another major consumer of energy. To overcome shortages of drinking water, some cities have desalinated water from the oceans. Such initiatives also carry a cost for the environment, e.g., raising salinity levels that impact upon the range of flora and fauna that live in coastal mud flats. All of these aspects of urban life increase energy consumption and greenhouse gas emissions.

Cities also create a micro-climate, usually being warmer and wetter than the surrounding countryside and this can also affect the flora and fauna of an area. However, cities have advantages such as concentrating opportunities and resources so that they can reach maximum numbers of people. Having tower-blocks to house large concentrations of people can make efficient use of space and reduce the costs of providing them with basic utilities such as electricity. Cities are also hives of creative activity in the arts, music and other forms of artistic expression. Moreover, they became and remain sites of class struggle whereby working-class people organized into trade unions, craft guilds, cooperative ventures and provident societies to overcome the worst excesses of capitalism and demand changes in workplace practices, legislation and policy that would enhance their quality of life (Thompson, 1963; Tholfsen, 1976).

In the West, the welfare state has helped to mitigate the risks of an individual being unemployed, disenfranchised and otherwise excluded. The notions of 'pooling risks' and utilizing 'social insurance' to cover the greatest number of people were useful ideologies in its creation (Webb, 1918). The solidarity implied in 'pooling risks' is being undermined by the current fiscal crises, although the ideology underpinning this attack on working-class lives has been around since the growth of neo-conservatism in the 1980s, under Thatcher in Britain and Reagan in the USA, where, arguably, this view has always held sway (Cloward and Piven, 1979). The safety net of a welfare state is usually absent in the Global South, and the family, especially its extended variants, and poor communities are called upon to provide for their own, regardless of their capacity to do so. Even countries where there had been a welfare state, particularly in the form of subsidies for education and health, had these destroyed by the Structural Adjustment Programmes (SAPs) demanded by the International Monetary Fund (IMF) and the World Bank as preconditions to provide loans to meet debt obligations, e.g., in Zimbabwe (Kaseke, 1996).

De-urbanizing the environment

Social and political movements have attempted to control the rate of urbanization and reduce its deleterious impact for a considerable period. Some began before the current 'green movement' and attracted professional elites. For example, in the UK, there was the Garden Cities movement that had sought to create living, self-sufficient cities with green spaces to provide recreational outlets for working people (Howard, 1902). Others created allotments to give landless urban workers the opportunity to grow their own fruit and vegetables (Crouch and Ward, 1997). The Federation of City Farms and Community Gardens was formed to encourage the creation of farms in cities to ensure that working-class children knew that animals that they might not otherwise encounter provided them with food, e.g., that cows produce milk.

One of the difficulties of these initiatives is that they yield isolated examples of worthy endeavours that created solutions for some people, but they did not provide a model that either could be rolled out across the whole of British society or could overcome the disadvantages of disparate urban settlements. For example, only Letchworth and Welwyn Garden City materialized under the Garden City movement. Even later, initiatives like the creation of a Green Belt around cities to control their spread into rural areas have had a patchy history and mixed successes. Designated Green Belt areas comprised 13 per cent of land in the UK in 2010. Their protection often depended on the party colour of the political class ruling in a specific local authority and their commitment to keeping these *in situ*. Conservative politicians were sympathetic to these guidelines being breached while Labour ones tended to enforce compliance. Sheffield City Council managed to retain much of the Green Belt around the city despite a number of central-government-sponsored policies that approved of their being partially dismantled for development purposes.

In the Global South, attempts to import models derived from the West have had patchy outcomes. The exchange of ideas about sustainable development in the context of urbanization has been useful in enabling communication and discussion about different ways of viewing the world between the diverse parties involved. Many ventures that imposed Western models of development on impoverished countries have been inappropriate and have caused more harm than good. Key among these have been the structural adjustment programmes (SAPs) that the IMF and World Bank (WB) enforced when agreeing to lend money to indebted low-income countries in the 1980s, especially those in Africa and Latin America. These international organizations received support, including funding, from Western donor governments and this led to the UN enforcing their policies as part of its remit in the social development arena. SAPs demanded the withdrawal of state subsidies from welfare and educational services in order to promote self-sufficiency and private-sector economic growth. The removal of the safety nets that people had in place increased poverty levels and social unrest, e.g., in Zimbabwe (Kaseke, 1996). At the same time, SAPs integrated these countries into a globalizing economy, the path of capitalist development and a neoliberal ideology. Visionaries like Jeremy Seabrook (2007) wrote about community initiatives that affirmed the construction of public housing to endorse the dignity of working-class people by ensuring that their environment was a habitable one. Seabrook was also a critic of rampant and uncontrolled urbanization as the key form of modern industrial development in British cities that has been exported elsewhere.

Civil society organizations, including overseas non-governmental organizations (NGOs) in the Global South, were also obliged to respond to capitalist imperatives, often providing limited funds that enabled people to set up local projects that mitigated some of the worst excesses of SAPs and provided a few individuals with well-paid jobs. Many of these NGOs moved on when donor funds disappeared (Doh and Teegen, 2003). Their departure left local people bereft of projects, resources and employment opportunities. These NGOs could also prevent the development of local alternative solutions to problems simply by being there at a time when the need arose. And their workers became trained in Western ways of thinking and doing, rather than investing in local knowledges and enhancing these (Shiva, 2003). Social workers were often employed by overseas NGOs. When they followed Western models of development, local residents often considered them part of the problem in their community development plans (Mohanty, 2003). When these practitioners supported the development of locality-specific and culturally relevant projects as the framework within which they would operate, even though their employer had stipulated the opposite, their results were more appropriate and successful (Edwards et al., 1999).

The welfare state that protected Western workers from some of the worst excesses of capitalist forms of urbanization did not just appear on the scene. It was the product of protracted struggles and inventiveness among working-class peoples in many countries. In the UK, waged workers created the trade union movement and welfare benefits through the guilds, credit unions and cooperatives to develop safety nets for their members as they sought to move away from reliance on the charitable benevolence of religious foundations or philanthropic individuals and towards the idea of rights to benefits and care during times of hardship (Thompson, 1963). In the Global South today, in the absence of substantive state inputs, international NGOs, civil society organizations, charismatic individuals and religious foundations are attempting to create structures that meet the needs of excluded groups. Recipients often subject these to trenchant critiques for not meeting their needs and for being culturally inappropriate (Mohanty, 2003). These criticisms draw upon traditional forms of caring and supporting others through times of hardship. These traditional ontological and epistemological bases can promote the development of alternative models of interventions and new knowledges. Below, I explore some of these initiatives.

Developing Alternatives to and within Centralized Urbanization

The report by the Conference Board of Canada (Brender et al., 2007) entitled *Mission Possible: Successful Canadian Cities* argued for the development of sustainable cities based on appropriate municipal governance structures, fiscal strength, and autonomy in pursuing prosperity. The authors identified the following ingredients as able to overcome the worst excess of urbanization:

- a strong knowledge economy to attract external investment;
- talented and skilled workers;
- a robust built infrastructure and transportation system capable of moving people and goods around efficiently and effectively;
- environmentally sustainable growth that followed sound planning and ecologically sensitive industrial principles;
- a socially cohesive society that includes affordable and attractive housing, access to low-carbon technologies for local consumers, a low crime rate, well-integrated immigrant communities, extensive cultural and entertainment amenities; and
- a strong social safety net.

These suggestions are enlightened. They are consistent with promoting publicly funded welfare states and other initiatives aimed at curbing the 'excesses' of capitalism and furthering mobility, including that of getting to and from work. However, they do not challenge the dominant model of market-driven social relations and economic growth.

My reading of these proposals is that such development aims to further industrial needs, while causing less social and ecological damage than if industry is left untrammelled to extract the maximum profit from existing resources – and not an alternative approach. Canadian cities like Toronto claim to follow this guidance, but, like other cities, it too has derelict areas and disadvantaged populations. Richard Florida (2004) echoes Brender et al.'s views by arguing that maintaining a high quality of life for all of a city's inhabitants is crucial to attracting a knowledgeable, entrepreneurial, creative and diverse workforce capable of innovating and pushing the frontiers that enable cities and countries to maintain their competitive economic advantage and address the problems of uncertain times. His analysis of the USA suggests that about one-third of its workforce belongs to the creative category – which includes scientists and engineers, as well as artists. Moreover, their impact can be aggrandized if they can be brought together as a coherent group working for the common good, instead of being high-flyers doing their individual thing. Having politicians daring to think big instead of limiting their vision to tax cuts and reducing services, he suggests, is more likely to produce alternatives to current urban thinking. Social workers can play important roles in getting local communities to engage in such discussions and lobby for more humane and ecologically sound alternatives than those offered by neoliberalism to improve the quality of life within an egalitarian framework that acknowledges environmental justice. There have been some attempts to develop such alternatives. I consider some of these below.

Microfinances and microenterprises

Mohamed Yunus was responsible for initiating the idea of microfinance to help some of the world's poorest populations in general, but women in particular, through what is now known as the Grameen (Village) Bank (Bornstein, 1996). In 1976, Yunus, then a professor and head of the Rural Economic Programme at the University of Chittagong in Bangladesh, encapsulated his thoughts in the concept of the Grameen Bank. A key notion behind the Grameen Bank's endeavours was that individuals would pool monetary resources to be able to borrow money later. It became an independent bank that relied on its borrowers to provide 90 per cent of its funds and operated according to the motto, 'Unity, Courage and Hard Work . . . in all walks of our lives'. The Grameen Bank enables people to maximize their capacity to raise cash when mainstream financial institutions refuse to lend them money because they perceive the risks of default as too high. By 2007, the Grameen Bank had lent US$6.38 billion to 7.4 million borrowers (Bornstein, 1996).

Those involved in the scheme could borrow small sums to start up a business that would enable them and their families to survive, and so was seen as making a significant contribution to poverty alleviation strategies. The bulk of those borrowing such money were women, who used their traditional skills for income generation purposes. Women receive 94 per cent of the loans made by the Grameen Bank, and access financial resources by using self-help measures based on the lending of money to each other. They are charged lower interest rates than those stipulated by mainstream banks and substantially less usurious rates than those demanded by loan sharks who typically lend to poor people. However, rates could be as high as 22 per cent. Microcredit schemes, thus, represent little more than market-oriented approaches to social problems, and, consequently, are consistent with the tenets of a neoliberal, globalizing world (Burkett, 2007).

Yunus' idea spread from Bangladesh to the rest of the planet. Business schools now promote microfinance initiatives and explore their theories in the classroom. This brainchild also earned Yunus the Nobel Peace Prize in 2006 and made him a millionaire. Social workers have been involved in both supporting women to form such groups and critiquing their weaknesses. Social work academics like Ingrid Burkett (2007) have researched microfinance projects and exposed fundamental failings in their operation. A key one is the gendered nature of their impact. Women, as the main users of these schemes, are unable to rise out of poverty, even if these can improve women's position somewhat. Another key reservation has been that such initiatives make poor women shoulder the debts of other poor women (Burkett, 2007).

Holding a group accountable for debts incurred by individuals who default on their payment ends up with poor women carrying the responsibility of making good such failures. As the entire group becomes responsible for ensuring that each person repays their loans, these schemes become collective ways of sharing risk that advantage those who raised the initial funds that enabled these women to borrow in the first place, if one of their numbers defaults for whatever reason. Their doing so protects the profit margins of those who provide knowledge and/or the resources that activate such schemes.

Other initiatives propose different ways of addressing poverty and providing women with opportunities to earn an income. For example, women community workers in India have worked with poor women in an organization called the Self-Employed Women's Association (SEWA) since 1972. It organizes poor self-employed women into a union that engages in self-help initiatives, aims to move women out of poverty by pooling resources and organizes collectively to fight against mistreatment. SEWA reinvests whatever surpluses are made in additional schemes that benefit women as a group. It has 1.3 million women members across the states of the Gujarat, Rajasthan, Bihar and Uttar Pradesh. It also supports women in Sri Lanka and Afghanistan. Their activities indicate a collective approach to poverty alleviation in contrast to the more individualistic one endorsed by the Grameen Bank. It engages in a wide range of income generation activities, and asks for tenders for many of these. SEWA also promotes energy-saving projects that reduce greenhouse gas emissions. For example, it has helped women access energy-efficient cooking stoves, and solar-powered lanterns. Social work educators support SEWA by sending students for placements and researching their activities.

Credit unions

Credit unions were local, working-class, democratically run fiscal initiatives that developed financial institutions that would lend money to members. They have an extensive history that began in industrial Britain to provide funds for working-class communities that were shunned by banks for loans because they lacked the assets that would protect their investments. Credit unions rely on individuals becoming members and saving, so that eventually they would be able to borrow money for large expenditures such as housing. The profits credit unions made were usually ploughed back into the organization to improve membership services. Credit unions are now found in ninety-seven countries throughout the world. However, they are not as common as they used to be in the UK. Community workers have been trying to revive them in some local communities, e.g., in Southampton and Durham, because they involve members in self-help initiatives and offer them loans at reasonable rates of interest.

Social enterprises

Another alternative response is provided by the social enterprise and cooperative movements. Social enterprises pool risks so that profits are not used to line the pockets of the few, but shared among the many to ensure that social concerns and care for the physical environment are integrated into economic development plans. Community workers and social work academics, e.g., Shragge and Fontane (2000) in Quebec in Canada, have argued strongly for the enhancement of such initiatives in local communities. A successful example they refer to is embodied by the Desjardines initiative which funds community development projects that other funders would ignore because they would deem the actuarial risk too great. This particular social enterprise has successfully challenged finance capital and grown from its small local beginnings as a credit union in Quebec 100 years ago to become Canada's fifth-largest financial institution. Desjardines operates as a cooperative network that lends money to local community groups seeking to improve the quality of their lives, microenterprises and social entrepreneurs aiming to build capacity in local areas. Desjardines typifies an alternative financial institution that has successfully challenged mainstream banking establishments. Nevertheless, while its endeavours have developed local communities, it has been unable to transform capitalist social relations and market-driven mechanisms into something else.

Another alternative approach to community needs is illustrated by Community Energy Scotland (CES). CES is a charity registered in Scotland and has been formed to expand community wealth, resilience and confidence by focusing on developing sustainable energy. For CES, sustainable energy development means the creation 'of projects designed to eliminate wasteful and inefficient energy use and generate energy from renewable means' in all parts of Scotland. It is part of a grassroots movement that utilizes renewable energy to generate employment opportunities locally, make more efficient use of energy and develop community-based alternative energy resources that can contribute to reducing greenhouse gas emissions.

Case study

Pedro was a young disabled man who lived in the Global South in a rural village of about 2,000 people. He had done well at the village school and was very keen to explore the new information technologies. Everyone in the village knew him, and many valued his help with computer problems. Some people, especially young children, made fun of him because he could not run and play with them, having contracted polio as a child. One year, a social work student called Antonio came into the village to do a placement and was able to get Pedro enrolled in distance-learning courses to enhance his computer skills and learn English. Pedro achieved very high grades for the work he did on this course. Several years later, when Pedro was nineteen years old, a representative from a large multinational firm came into the village looking for people who could work in his company that was going to be based in a town about 50 kilometres away.

Pedro had always wanted to work away from the village, where he could only get unpaid work helping out his family, neighbours and friends. So when one of his friends told him that there was a man in town looking for computer experts to employ and gave him the man's card, Pedro contacted him by email, giving him details about his course and experience and asked for a job. Eventually, Pedro received a reply that invited him to go to the town to be interviewed. Getting there was a challenge for him, given that there was only a gravel road out of the village and no bus. Later, a farmer who took vegetables to market offered him a lift in his cart. Pedro was delighted and took great care not to dirty his only suit before the interview. The farmer also promised to take him back to the village at the end of the day if he had finished his interview.

Pedro did not like the look of disdain on people's faces as he entered the new headquarters of the company. This did not improve as he sat in a room with others waiting for his turn to be called. His determination to persevere and be one of the ones to succeed paid off and Pedro was offered a job. But, to his surprise, he was appointed to a call centre to answer telephone calls mainly from customers based overseas and who spoke English. Nonetheless, he accepted the offer and set about organizing his move from the village to the town. He had hoped that his new employers would help him find housing and settle, but this was not forthcoming. The money that he was earning did not allow him to rent anywhere in the nicer parts of town. Eventually, he found a room about one hour away from the office via several buses and two long walks. It was in a run-down tenement where there was one toilet and one shower shared between twenty people. The room smelt of mould, and Pedro could not seem to get rid of it although he left the one window open. Soon his clothes, few as they were, started smelling of mould and his colleagues at work started complaining about the way he smelt. At home, the other tenants pushed and shoved him as they sought to get round him on the stairs, and called him names because he was too slow. Pedro felt lonely and isolated and soon became depressed. He did not eat or sleep much either.

One day, he was in a cheap coffee shop near his home drinking coffee when someone came and sat next to him. It was Antonio, who was now working for an NGO that had been sent there to help residents improve their lives. Pedro started laughing and found it difficult to stop. When he finally did, he asked Antonio if he were joking. What could he do in this god-forsaken area? Could he get him a job that paid him enough to rent a decent home? Could he stop people from calling him names and hitting him because he had deformed legs? To his surprise, Antonio said, 'Yes'.

The next day, Antonio visited his tenement and began to talk to all the tenants. He made copious notes of all their complaints and the lack of repairs throughout the building. He found out that the absentee landlord was none other than a local subsidiary of the multinational company that Pedro worked for. Having collected this information, Antonio called a meeting of the tenants so that they could all decide what they wanted to do and what actions they wanted to prioritize. Repairing and modernizing the building, installing more toilets and showers, and reducing rents were their top three priorities. After that, they wanted better and cheaper transportation. They then discussed what they should do and who should do it. In the process, they discovered that, at the end of their long journeys, most workers were employed in places within a 10-minute walk of each other. Antonio worked out that they could pool their money together to hire a bus to take them to and from work for a fraction of the cost they currently paid separately, and that it would cut their journey time to about 20 minutes and reduce carbon emissions. The tenants unanimously voted to hire a bus to take them to work. Antonio also assisted Pedro in acquiring better prosthetics which enhanced his mobility and confidence, to the extent that he became vocal in meetings and active in implementing agreed actions.

Antonio helped the tenants form a tenants' association, secure free legal assistance to contact the landlord, compile a dossier of their complaints and present their demands. Progress on the repairs is slow, but some work has begun and the tenants are maintaining their pressure on the owner of the building. And, under Pedro's guidance, they turned an overgrown rubbish dump that had been the yard into a garden with a vegetable plot and a pond for local amphibians and fish. Pedro was astounded to be elected Chair of the Tenants' Association, and became delighted with having a role that took him out of himself, and finally to be valued by others. The nightmare of living in an urban environment began to clear, although he knew it would always be hovering in the background. And, with Antonio's guidance, he learnt new skills and gained confidence in his abilities. His interest in enhancing his environment – social and physical – expanded into ensuring that the landlord put energy-efficient appliances and water-saving toilets into the tenement. And, he asked to be promoted at work.

Pedro's story indicates how small interventions by social workers in the daily lives of marginalized people can have profound effects in initiating individual and collective changes that enhance the quality of both their lives and their communal environments. It also demonstrates that green social work has to be holistic and provide training that facilitates the right of people to be cared by, and care for, others, while protecting the environment, if it is to go beyond the mainstream recommendations for future policy on disaster interventions which currently focus largely on: early evacuation of individuals with special needs; improved training of staff to include practice drills; better communication systems; and increased funding of the local state and health departments (Kirkpatrick and Bryan, 2007).

Conclusions

Urbanization is a global phenomenon that is likely to increase. However, the current neoliberal capitalist approach to it is imposing considerable costs on people and the Earth's physical environment, including its flora and fauna. The demands of urbanization for transportation, land and other resources like water, fuel and time need to be subjected to scrutiny and interrogated for their relevance to an increasingly heavily populated world that has scarce physical resources and is losing much of its biodiversity. Social workers have crucial roles to play in: raising questions about an equitable sharing of the planet's resources; engaging in sustainable development when mobilizing local communities; ensuring that locally relevant and culturally appropriate strategies are put in place to respect people, living things that share their habitats and the physical environment in their communities; and developing sensitive, holistic and sustainable approaches to meeting the needs of people and the planet.

4

Industrial Pollution, Environmental Degradation and People's Resilience

Introduction

Very serious accidents have occurred when human controls over the products that scientists have invented have gone awry. Examples of these have been the leakage of dioxins or radioactive materials in many spots throughout the world, ranging from India to Italy and the USA. The ensuing disasters have gravely undermined people's health and well-being, defined as the rights to enjoy the products of the Earth safely and to develop their skills and talents to the fullest extent, as articulated under Articles 22 to 27 of the Universal Declaration of Human Rights (UDHR). Industrialization has produced particular forms of pollution that have had deleterious consequences for people's health. Asthma, the rise of various respiratory ailments in many countries, and a range of disabilities can be traced back to the lack of controls on pollutants that industrial processes discharge into the atmosphere, oceans and land. The banning of lead as an ingredient in petrol used to drive motor vehicles and the banning of DDT in pesticides are well-known outcomes for people demanding the protection of their health from the effects of toxic chemicals. Such successes indicate how small behavioural changes in everyday life activities can have beneficial results for others.

In this chapter, I examine the impact of industrial disasters on people's lives and their capacity to develop resilience, expressed as the capacity to deal with unforeseen events that undermine their normal routines and ways of being, like those that occurred at Three Mile Island in the USA, Chernobyl in the Ukraine, and Bhopal in India. Poor health and higher morbidity arising from increased rates of cancer, rising levels of congenital disabilities and the loss of livelihoods, social isolation and stigma are some of victim-survivors' concerns that social workers assist with following industrial disasters. I highlight the importance of existing social resources, networks of responsibility and legal instruments that can hold perpetrators accountable for their actions, while exploring social workers' roles in developing resilient responses to industrial disasters. This work can entail action locally, nationally, regionally and internationally. Thus, social workers, including those acting as community workers, have to acquire knowledge, resources and allies at all these levels. Crucial in their knowledge repertoire is being familiar with the concepts of vulnerability and resilience, risk assessment and risk reduction, impact assessments, risk mitigation, adaptation and coping capacities. Strength-based approaches can be crucial in reducing vulnerabilities and enhancing resilience.

Reconceptualizing Resilience

Determining people's capacity to overcome adversity and move on with their lives is central to notions of human, social and environmental development that are sustainable and resilient in the short, medium and long terms. Dealing with industrial accidents or natural disasters requires forms of resilience that transcend everyday responses, i.e., robust resilience. Resilience is a concept that social workers use in mobilizing individual and community reactions to calamitous events. Resilience as a concept began in the physical sciences, where it was defined as the capacity of materials to respond to stress, and moved to the social sciences, arts and humanities, where the concept became a system management tool for controlling and containing crises (Manyena, 2006). Used in a managerialist and linear manner, resilience loses its capacity to promote sustainability involving a step-change that develops resilient resources for people, communities, institutions and organisms to survive or thrive under conditions of unremitting and/or threatening change.

Critiques of this managerial usage have refocused resilience as an active concept developing the capacity of systems, whether natural, human or hybrid, to sustain themselves when facing endogenous and exogenous shocks. Moreover, resilience has been reconceptualized as an emergent property with non-linear and fractured characteristics in that a system could be resilient along one dimension but not another, and might lose resilient structures over time. Expressed in these terms, resilience becomes an emerging property with the potential to rebalance and refocus the idea of managing system shocks, to enable the development of robust resilience prior to, during and after a disaster. Robust resilience, unlike mundane or everyday resilience, involves preventative measures, crisis responses and long-term reconstruction and is essential in tackling structural inequalities (Dominelli, 2012).

Social workers suggest that defining resilience in a non-linear way is crucial in systems that facilitate people's actions as agents who evaluate several dimensions of risk simultaneously and intuitively, before choosing a particular course of action. They also envisage resilience operating on a continuum that ranges from non-response, which can lead to a system failure, or even turn out to be resilient in some cases; to survival through adaptation, accommodation or making do with a situation; and on to thriving responses that are innovative and transformative. Resilience responses can produce structural changes in systems and innovative thinking that fees situations to develop into something new and different. I have tried to articulate this nuanced complexity in figure 4.1.

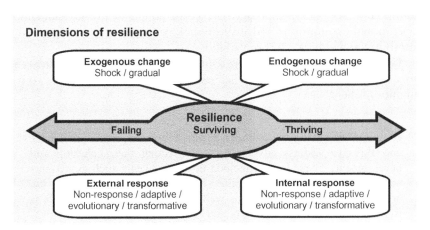

Source: Dominelli (2012)

Figure 4.1   Dimensions of resilience chart

Klein et al. (2004) argue that the lack of agreed definitions about the concept of resilience and the failure of experts to operationalize it effectively for practice are major drawbacks in using it to guide interventions that reduce risk. Social workers are among those groups who refer to resilience conceptually without operationalizing it to any significant extent. Such fuzziness in thinking can be reduced by adopting Klein et al.'s suggestion of seeing resilience as one of a number of factors that enhance adaptive capacity, which they define as attributes that enable a system to self-organize and absorb disturbance whilst remaining in the same state. This, Klein et al. (2004: 15) argue, assists practitioners in 'mov[ing] away from disaster recovery to hazard prediction, disaster prevention and preparedness'. It can also assist the development of robust resilience.

Resilience can be reactive or proactive (Dovers and Handmer, 1992). In reactive forms of resilience, people focus on the adaptive mode, with survival being their main preoccupation. They plan for the future by strengthening the system's existing capacity to resist change and maintain the status quo. However, unintended consequences and/or further shocks to the system may alter both this objective and its outcome. Adaptive systems tend to be more sensitive to political imperatives and are more inflexible. By prioritizing stability over other concerns, adaptive approaches can endanger the future viability of the system. This can be seen in politicians' reluctance to tackle polluting firms, often by dragging their feet and being tardy in enforcing legislative stipulations. Under conditions of proactive resilience, people consider change inevitable and seek to develop systems that can readily adapt to new demands and situations. The more flexible, proactive response adapts by quickly changing operational assumptions and institutional arrangements to reduce vulnerability and enhance future viability.

Gubbins (2010: 8), considering a more collective response undertaken by communities which discuss such issues with each other, defines community resilience as:

> communities having the confidence, capability, resources, knowledge and skills to address adverse factors affecting their cohesion and development. These factors include dramatic events such as extreme weather events; energy cost spikes; blackouts and energy insecurity; and national financial crises; as well as more chronic issues such as rural depopulation; fuel poverty; ageing communities; urban deprivation and unemployment.

This definition suggests that resilience has implications for every aspect of life, and is a view with which social workers would concur. Resilient responses can also provide a cushion of certainty in uncertain conditions. Additionally, social workers' preoccupation with resilience would also focus on understanding the links between it and vulnerability.

Defining vulnerability

Vulnerability is the flip-side of the resilience coin. In climate change debates, the UN (2006) uses the definitions of vulnerability articulated by the Intergovernmental Panel on Climate Change (IPCC). It is:

> The degree to which a system is susceptible to, and unable to cope with, adverse effects of climate change, including climate vulnerability and extremes. Vulnerability is a function of the character, magnitude, and rate of climate change and variation to which a system is exposed, its sensitivity and its adaptive capacity (IPCC quoted in UN, 2006: 11)

The UN has also drawn on the ISDR (International Strategy for Disaster Reduction) definition of vulnerability as:

> The conditions determined by physical, social, economic and environmental factors or processes, which increase the susceptibility of a community to the impact of hazards. (ISDR quoted in UN, 2006: 11)

Vulnerability can be experienced as a result of endogenous (internal) or exogenous (external) factors. The degree of vulnerability depends on a system's exposure to hazards, its resistance to these and its resilience in coping with them. Understanding the sources of vulnerability, which are identified as reposing in physical, social, economic and environmental factors, enables experts who intervene in such situations to mitigate risk, make firmer predictions about likely risk and suggest what can be done to reduce it. Risk reduction involves professionals in enhancing people's capacity to tackle the hazards that they might encounter. In disaster situations, degradation of the environment, including the contamination of land, air and water; reductions in biodiversity; and people's movements can increase existing vulnerabilities. These have to be reduced to enhance resilience. Such endeavours can benefit from multidisciplinary approaches to the work. In achieving this goal in disaster situations, emergency response teams, including social workers, seek to determine susceptibilities in the systems involved holistically.

To build resilience, residents can involve: engineers to assess infrastructures including water supplies, power grids, sanitation systems, communications and transportation systems; sociologists to explore how resource allocations impact upon social stratification systems, according to social divisions like ethnicity, gender, age and ability; economists to consider resource availability and the economic bases from which reconstruction and recovery could begin; health and social care practitioners in examining health provisions and social care services; educationalists to determine and develop literacy levels that promote active engagement in community decisions; community development workers to focus on governance systems, people's access to human rights and social justice; and social workers to assess peoples' needs and ensure that these are met, and encourage organizations to provide locality-relevant assistance that upholds cultural values and traditions. This will facilitate the development of co-produced solutions that engage interdisciplinary insights with residents' knowledge leading to collective action and robust resilience at the level of community. Moreover, engaging communities in this way encourages their ownership of the solutions that are devised.

Social workers undertake risk assessments and conduct impact assessments among affected populations to calculate both vulnerability and risk. The social work literature tends to emphasize endogenous vulnerabilities. This approach focuses on individual (personal or system) characteristics and can heighten the likelihood of people being blamed for their vulnerabilities and being held accountable for a failure to adapt. Minimizing vulnerabilities is central to avoiding and reducing risk (Swift and Callahan, 2009). Assessing vulnerabilities in contexts of uncertainty is difficult (Adger and Vincent, 2005). To develop resilience within communities effectively from a green social work perspective, social workers cannot treat the Earth and its bounty as a means to an end. The Earth's treasures have to be valued in and of themselves, as well as for mattering to human beings.

A group of ecological scientists organized the Resilience Alliance to undertake research into resilience and sustainable development and prepare a paper for the World Summit on Sustainable Development in Johannesburg in 2002 (Folke et al., 2002). For them, resilient cities that focused on developing system function and working as learning organizations were better prepared for hazardous events and able to cope with their consequences more effectively (Klein et al., 2004).

Case study

Jackie was a community development worker supporting residents living near a flood plain in resisting its development for luxury apartments. The location overlooked a river that already had a pub on its banks. The community acknowledges the lack of local housing, but it wanted these to be affordable for the young people in the locality and built on a brown-field site instead. They wanted to preserve the water meadows and felt that the pub, popular among locals and visitors from outside the area, was sufficient development for that locality.

The company that proposed to build the apartment complex was based overseas and its local representative had no authority to make decisions on its behalf. The community was determined to stop the development, and with Jackie's help sought a holistic analysis of the flood plain, its flora and fauna, and the costs of the luxury development during its construction phase and into the future. They also included forgone opportunities in the form of recreational facilities and money that could be used to provide housing that the community actually needed. Using a range of external experts identified by Jackie, the community prepared a lengthy dossier and asked to meet with the company's director to discuss their alternative plan which proposed a combination of social housing and luxury apartments on a brown-field site that overlooked the river but would not damage the flood plain because infrastructures such as housing, roads, communications and public utilities could utilize land already under human use.

This case study exposes the importance of a holistic approach that takes into account the needs of all stakeholders rather than privileging the interests of the monied few; takes into account current and future needs regarding infrastructural developments like housing and the services that have to be provided for those who will go to live in these; considers the environment as a living entity, not just a commodity with a price tag attached; and brings together a variety of experts and local people to exchange knowledge and produce more environmentally sound solutions to problems.

Industrial Pollution and Accidents

Industrial processes have by-products that are not needed after the production of particular goods. These are usually discharged into the atmosphere, the Earth's rivers, lakes and oceans, or even land surrounding a factory, often with little regard for the people, flora or fauna living nearby. They are termed pollutants because they are substances that have a detrimental impact on human beings, plants and animals. Pollutants cause physical illness that impairs people's health, like asthma and a range of respiratory diseases. Chemical pollutants include asbestos fibres that lead to pneumonocosis; substances that cause physical deformities including congenital ones; industrial products, including neon-lighting that leads to migraine; and others that result in a range of illnesses captured by the phrase 'sick building syndrome' (Murphy, 2006). The discharge of chemicals into river systems has led to loss of fish and other aquatic life, e.g., salmon in the River Thames (TRRT, 2011). Some chemicals that formerly have been utilized routinely in the West and continue to be used in daily life to exterminate pests in crops or malaria-causing mosquitoes in the Global South, e.g., DDT (dichlorodiphenyl-trichloroethane), have caused breathing difficulties and irritated sensitive body organs (Carson, 1962). DDT is now banned in Western countries. Meanwhile, chemicals from the phosphate fertilizers that raise food yields can seep into aquifers and render water unfit for human consumption. The pollution of water in rivers and lakes through chemicals in fertilizers and other products is known as eutrophication (Connett, 2003). Waste products from their production also escape into the air and damage the ecosystem including the land, plants and animals in what has been termed 'fluoride poisoning'.

Industrial accidents can be caused by human error and/or technological malfunction. Whatever the reason, suffering to people, and damage to plants, animals and the physical environments can be severe. In some situations, remedies are possible. In others, nothing can be done and the site has to be put under lock and key, with access restricted for substantial periods. I now examine several well-known incidents. Although they have happened in the past, these events remain highly controversial. Disputes can be about who or what caused the accident; whether particular health problems, including cancers, congenital deformities and respiratory diseases, can be attributed to a particular accident; who should pay for the damages resulting from it; how much money should change hands; and what precautionary measures have to be adopted in future if the manufacture of the products in question is to continue, as it often does (see www.lenntech.com/environmental-disasters.htm).

The tragedy of the Bhopal gas explosion

Bhopal in India was the site of one of the world's worst chemical disasters. It took place in 1984 when methyl isocyanate gas escaped from the Union Carbide India Limited (UCIL) plant making the pesticide Sevin. UCIL was a subsidiary of the Union Carbide Corporation (UCC), an American multinational, which was taken over by the Dow Chemical Company in 2001 (Jackson, 1993). The ensuing explosion released a plume of dangerous chemicals that killed and maimed thousands of people and animals, and destroyed much vegetation in the area. The escaping gases polluted the atmosphere and chemicals in storage pools contaminated groundwater and river waters causing the loss of millions of fish and other aquatic life. Although the number of deaths and people whose health was compromised was substantial, the actual figures are contested. Some victim-survivor claims for compensation are outstanding, and demands for the executives to be held to account have been frustrated. Several civil and criminal cases are pending in both the USA and India, as the survivors of the disaster and/or their descendants struggle to get recognition of and compensation for their losses. Meanwhile, several UCIL employees were found guilty of negligence by a court in Bhopal, with a two-year sentence imposed on several local managers as recently as 2011. Litigation and the resolution of issues at stake for those involved in the Bhopal disaster were complicated by the sale of the subsidiary company in India after the accident and the headquarters of the parent company being based in the USA.

Bhopal narratives allege that cuts in staffing levels, reduced frequency in carrying out safety procedures, cross-cultural barriers in communication, poor management and inadequate maintenance of the facilities before the disaster contributed to it. Additionally, these assert that continued poor management, neglect and inadequate maintenance of the plant and surrounds since the disaster have aggravated its deleterious effects. Its location in a heavily populated area with slums in the vicinity meant more people were affected. Lack of accountability, poor disaster plans and inadequate health and social care services to respond to the high levels of need resulting from the accident compounded the difficulties that local residents faced in dealing with the disaster's consequences for their community and physical environment (Gottschalk, 1993).

As a result of the lessons learnt from the Bhopal disaster, the chemical industry launched a voluntary initiative called Responsible Care, to improve health, safety and environmental performance in whatever locations chemical producers have major plants. Responsible Care can be found in the fifty-two countries that manufacture 90 per cent of the world's chemical production. The initiative also promotes dialogue with local communities and their governments. Thus, Responsible Care provides leverage that social workers can use when working with community groups to enforce safety standards. Social workers have supported the people of Bhopal, immediately after the disaster and since by providing psychosocial support and helping them to access medical care and supplies, and material resources. Supporting people in dealing with the traumas caused by the loss of life, livelihoods and homes; lack of response from senior company executives in the company and politicians at various levels; and subsequent litigation are all important forms of support for people through disasters and beyond.

Social workers could play even bigger roles in supporting local peoples' demands to hold multinational firms accountable for their decisions and for these companies to fulfil their duty of care towards employees and the environments in which firms are located. Some of these efforts would be preventative and include consciousness-raising to enable people to become aware of their right to protection from the risks that the production of certain chemicals in their locality holds for their health, in the event of an accident, and how to evacuate the area as quickly as possible should one occur. Social workers should note that their taking an active role in demanding accountability from powerful multinational companies or their executives could jeopardize their position with their employers, and may risk their employability if these powerful others complain about their behaviour.

Chemical Pollutants and Radioactive Gas Escapes

Toxic chemical pollutants and discharges

Dioxins are toxic chemicals that can produce disabilities among otherwise healthy people and increase the number of congenital deformities in affected populations, and compromise plant and animal health. For example, the leakage of dioxins and other chemicals produced by the Coalite Company near the river Doe Lea near Chesterfield in the UK during the 1990s highlighted the toxic nature of these chemicals, and their impact on the health of people and animals, including the rendering of cows' milk unfit for human consumption (Jones and Duarte-Davidson, 1997). The company went into liquidation and defaulted on pension payments owed to its employees. Now, the Avenue Coking Works at Wingerworth is decommissioning the site. In 2011, local residents like Michael Holdaway continued to complain about the effects of gases being released upon family health (www.transitionchesterfield.org.uk/content/smell-avenue-works).

In 1976 in Seveso, Italy, dioxins were discharged into the atmosphere and affected the health status of people living in the locality when 2,3,7–8 tetrachlorodibenzo-p-dioxin (TCDD), a component of Agent Orange used earlier in Vietnam, fell on the town. The TCDD was emitted by a factory in nearby Meda which was owned by subsidiary companies that could be traced back to the parent Roche Group. The emission led to the evacuation of people from the most heavily affected areas, and caused soil contamination which affected people, plants and animals. However, the people were unaware of the presence of TCDD in their environment. Fear caused a delay in initiating emergency plans including the evacuation of people, once the danger became known. Decontamination cost nearly 120 billion liras. The clean-up exercise became a cause of disquiet because it was poorly organized and executed (Bertazzi, 1991). The outcry around this accident compelled the European Union to proclaim guidelines, termed the Seveso II Directive, to improve safety at industrial sites, in 1996.

Social workers comforted those evacuated and fearing risks to their health and provided psycho-social counselling. They could have played a much larger role by: making people aware of the potential dangers in their situations beforehand, supporting an earlier evacuation of those affected, and lobbying for higher and enforceable safety standards. There are also clear advocacy roles that social workers can undertake to support Chesterfield residents like Michael Holdaway.

Three Mile Island, the USA

A serious problem in a nuclear reactor on Three Mile Island in the USA resulted in the release of radiation in 1979. Despite earlier warnings that an accident could happen, remedial measures were not taken. The ensuing damage to Unit 2 was caused by a combination of human error and mechanical failures that caused radioactive gases and iodine 131 to enter the atmosphere when Unit 2 lost coolant and had a partial core meltdown that was rated 5 on the 7-point International Nuclear Event Scale (INES). Confusion over exactly what was happening caused by an 'unanticipated interaction of multiple failures in complex systems', delayed evacuation plans. The discharge of substantial amounts of radio-active waste water into the Susquehanna river, authorized by the Nuclear Regulatory Commission (NRC), led to a loss of credibility for the agency and its attempts to control the crisis. Clean-up costing US$1 billion started shortly after the disaster and lasted until 1993.

Pennsylvania State was also poorly prepared to respond to the disaster, having not stocked potassium iodide which people have to take to reduce the impact of radioactive iodine on the thyroid. Although the Kemeny Commission reported no ill-effects from this event, this finding was disputed in a report produced by J. Samuel Walker (2004). The Three Mile Island accident highlighted the failure of safety mechanisms to prevent the disaster, and the need for extensive research into controlling radiation fallout, de-radiating or reducing the half-lives of the isotopes released as part of the radiation, and finding cures for the cancers that can result. The accident led to improved training for control room operators in nuclear plants, better emergency evacuation plans within a 10-mile radius of Emergency Planning Zones, and rehearsals to prepare for actual disasters. This accident caused the cancellation of some nuclear plants, although Unit 1 in Three Mile Island was allowed to resume activities in 1985. In 2009, it received permission to continue operating until 2034 even though residents living in its environs voted for it to be decommissioned. This response depicts the failure of politicians to adhere to democratic decisions taken by the electorate. Social workers could help people mobilize to enforce their vote and compile evidence needed to uphold their views and present information to ruling elites.

Chernobyl, Ukraine

A major nuclear disaster involving an explosion and a fire in reactor number 4 occurred in 1986 in Chernobyl in the Ukraine, then a part of the Soviet Union, as operatives were attempting to test safety procedures in the event of a power failure. Evacuation was slow, and information about the disaster was not easily forthcoming. The Soviets paid 18 billion roubles to contain the damage. Although the figures are contested, the incident is believed to have caused the immediate deaths of forty-seven workers exposed to radiation, and a rise in cancers related to it. The accident was classified as a Level 7 Event and released radioactive substances into the atmosphere that covered Western Europe and other parts of the Eurasian continent. The area immediately around the plant is still not habitable. Like other nuclear disasters, news about the extent of the damage and the safeguards that were in place was limited, making it difficult for people to decide what steps they should take to mitigate its effects. Nonetheless, this accident led to a decline in the popularity of nuclear power as a trusted energy source (Englund, 2011). Supporting people living in the shadow of a nuclear plant provides an opportunity for social workers to assist local people in obtaining data about its operations and building their resilience to future potential disasters.

Chalk River's nuclear accident

The nuclear facility in Chalk River, Canada is one of a few firms that manufacture radioactive isotopes for medical purposes. It had radioactive leaks on a number of occasions between 1952 and 2009. The facility is old and repairs would cost CDN$1 billion. The federal government shelved plans to replace two reactors in 2008. The federal government is reluctant to fund repairs directly, although it is pursuing options for a private–public partnership. The plant's anticipated lifespan until 2016 has been extended to 2021. The lack of clear plans for its continued survival and a proposed restructuring of its parent company, Atomic Energy of Canada Limited (AECL) have created uncertainty among the scientific community employed at Chalk River and local residents. If these scientists seek posts elsewhere, the loss of expertise and intellectual and productive capacity would be difficult to replace (Paperny, 2011).

Cleaning-up nuclear accidents can also cause long-term damage. Canadian veterans involved in the Chalk River clean-ups of the 1950s have become seriously ill, many with cancers. Some are seeking to establish this as fact and secure compensation from their government. The Canadian Coalition for Nuclear Responsibility supports veterans such as B. H. Paulson, who clean up atomic disasters. He seeks compensation for alpha-ray radiation that he alleges caused his cancers when the Canadian RAF (Royal Air Force) sent him to help dispose of the waste from one of the accidents that occurred at Chalk River. He claims that his protective gear worked only against gamma radiation, not alpha radiation (www.ccnr.org/paulson_legacy.html). Mr Paulson's application for compensation was rejected. He became disillusioned when informed that there was no record of his being in Chalk River, although he was one of 600 soldiers sent to clean up the debris and secure the facility. Social workers could support him during this ordeal and assist in his struggle for compensation.

Fukoshima Daiichi, Japan

The nuclear explosion in Japan in 2011 represents a multiple hazard situation in which an earthquake that reached a magnitude of point 9 on the Richter Scale, and a subsequent tsunami that breached a retaining seawall, flooded the entire Fukoshima Daiichi plant, destroyed the plant's connection to the power grid and damaged the technology that was to ensure that the reactors remained cooled. The owners of this nuclear plant failed to control and prevent radiation leakages. The World Bank estimated the damage caused at US$235 billion. Three reactors experienced full meltdown, a fourth was damaged by fires and the fifth and sixth were partially damaged. People were eventually prevented from remaining in the surrounding area, and food grown in its precincts was banned from sale.

The government's response was considered inadequate for failing to grasp the enormity of the disaster and not initiating the appropriate responses quickly enough. Prime Minister Naoto Kan resigned over his handling of the affair. The event was initially ranked 5 on the INES, despite alternative views held by international experts. It is currently ranked 7 which is the highest level on this scale, and on a par with Chernobyl. The private company that owned the facility, the TEPCO (Tokyo Electric Power Company), did not release the information that people needed to take protective action. Workers who were trying to control the blaze in the nuclear reactors found their health compromised. TEPCO also had not responded to earlier concerns about the safety of the plant raised by experts within and outside Japan, such as those at the IAEA (International Atomic Energy Agency) (Fackler, 2011). The entire site is to be decommissioned when it is safe to do so and nineteen of the country's nuclear reactors have been shut for safety reasons (McCurry, 2011). TEPCO's reactions raise issues about the accountability of private firms to people living in the communities that are affected by the decisions they take or do not take. Social workers can help local communities obtain information about such concerns and act upon it.

The fallout from Japan's disaster was wide-ranging. Many people who had reservations about the safety of nuclear energy had their fears confirmed, and opposition to the expansion of this energy source stiffened. In countries like Germany, the government decided to halt the sector's expansion and close down existing facilities as soon as convenient and safe to do so. Germany has refused to endorse the building of more nuclear reactors in the future, focusing instead on renewable energy supplies (Dempsey and Ewing, 2011). Social workers have been involved in public demonstrations to support these views and avoid energy disasters. Working with physical scientists to find solutions that are people and environmentally friendly is also a crucial part of the social work disaster repertoire.

Caring for the Carers in Disaster Situations

In the 2011 disaster in Japan, local social workers, including educators and students were overwhelmed by the level and range of demands resulting from extensive damage caused by multiple hazards. Psychosocial interventions are likely to last some time, as are the rebuilding endeavours. People evacuated from the site are likely to require permanent housing elsewhere as the radiation threat will last some time. Social workers are helping people address these issues. Experiences in Chile, Japan and Aotearoa/New Zealand indicate that they can become stressed, burnt-out and unable to cope without supportive supervision (personal communications) and good self-care. Cronin et al. (2007) argue strongly that preventing burn-out among the helpers is insufficiently addressed in disaster responses, when caring for their mental health needs is essential. Virtual helplines like the one described below can support practitioners and other victim-survivors to help debrief them at their convenience.

Virtual support for carers and victim-survivors

The residents of Christchurch, Aotearoa/New Zealand, were struck by two earthquakes. One occurred in late 2010, the other in early 2011. The second 'quake exacerbated local conditions because people had yet to recover fully from the previous one. The extensive devastation caused put huge pressures on local emergency services, although assistance was given by the rest of the country and many people from overseas. Local emergency workers, including social workers, had difficulty working virtually round the clock for days on end without breaks, especially in situations where the victim-survivors were traumatized by these events (personal communication). A call for relief for overworked social workers was received by the Chair of the Disaster Intervention Committee of the International Association of Schools of Social Work (IASSW) and a response was set in train. Given limited resources and distances from the 'quake scene of many volunteer social workers who responded to the request, it was decided to set up the Christchurch Virtual Helpline as the support vehicle. Victim-survivors who wanted to talk through their experiences with others were given the choice of phoning, or using email or SKYPE in order to conduct their conversations. This response highlighted the resilience of social work professionals in using modern technologies to cross the miles in support of victim-survivors who included residents and over-stretched professionals. It also ensured that infrastructures and resources were not stressed by additional demands from overseas helpers. Undertaken under the auspices of the Disaster Intervention Committee, the qualified and experienced social workers who responded were committed to using the guidelines given below.

Guidelines for a Virtual Helpline Support Network in Disaster Situations: The Christchurch Example

Acting ethically in disaster situations is crucial. As social workers, we each have our own national code of ethics to refer to. The IASSW-IFSW Ethics Document (www.iassw-aiets.org), and IASC Guidelines for Psychosocial Interventions (www.who.int/mental_health/emergencies/ guidelines_iasc_mental_health_psychosocial_june_2007.pdf), and guidelines provided by the Telephone Helplines Association (www.ncvo-vol. org.uk/sites/default/files/document/GoodTelephonyGuide.pdf) are also useful in doing this work. Our main role is to listen to people and to enable them to tell their stories to someone who will not take advantage of their vulnerabilities, but will know to whom to refer the caller if additional help is required. Adhering to principles like those provided below is essential to acting ethically in such situations.

Principles to follow in providing online or telephone support

- *Do no harm*
- *Supportive listening*
  Hearing what people have to say and validating their experiences
- *Culturally relevant and appropriate responses*
  Gain an understanding of the culture of the people who will call, noting that there will be variations within Aotearoa/New Zealand. A useful website to begin with is:

www.newzealand.com/travel/about-nz/culture/culture-home.cfm
- *Signposts to other resources*:
  www.msd.govt.nz/emergency/index.html
  http://canterburyearthquake.org.nz/

As you are likely to be from a different culture, ensure that you ask the caller about their particular cultural needs and respond appropriately.

- *Issues – Feeling unsafe, dealing with uncertainty*
  Encouraging people to consider the elements which cause them to feel safe, by looking for the certainties that they already have in the midst of uncertainty and hardship, may help here.
- *Issues – Confidentiality*
  Confidentiality is important to maintain, and the limits to which this confidentiality exists should be clarified at the outset. This makes it 'contingent confidentiality'. You will have to report a threat of harm to oneself or others and have some idea of whom to refer such individuals to should this need arise.
  www.police.govt.nz/
  http://canterbury.webhealth.co.nz/provider/service/view/2035035/
  www.lifelinechch.org.nz/pes.html
- *Survivors not Victims*
  Consider people as survivors, not victims, so help them look for their strengths and how they can use local resources to help themselves after your discussion by SKYPE, phone or email. Note that it is easy to become entrenched in the minutiae of the experience and in repetitive details that can, at some point, become unhelpful to the recovery process. It may be useful to establish time limits to the discussion at the outset while remaining open to further conversations at another convenient time.
- *Caring for Oneself*
  You may need to 'debrief' yourself either in your locality or through the other members of the Helpline. Make sure you have these arrangements in place before you begin your discussion with disaster victim-survivors.

It is essential that helpers act ethically at all times.

Notes

1. Please note time differences between your country and Aotearoa/ New Zealand.
2. Please offer to pay the costs of the telephone call as part of your contribution to the relief effort. This includes 'collect' calls from

Aotearoa/New Zealand. If you do not wish to pay for these, please let the Head of the Disaster Intervention Committee know that you will be accepting only email or SKYPE calls and this information will be placed on the database.

3. You should keep brief notes of your phone calls or discussions in case you need to refer to these and be prepared to share these with other members of the Helpline if necessary. Make sure you clarify such practices with the caller and obtain their consent to do so at the beginning of the call.

4. If you have colleagues who wish to join, please have them contact the Head of the Disaster Intervention Committee.

Social Worker Involvement in Industrial Pollution and Environmental Degradation Issues

Zapf (2009) argues that social workers are well placed to engage in environmental issues, especially as they hold the 'person-in-the-environment' as a basic tenet of practice and could play a key role in the formation of an 'environmental state'. To do this, Zapf (2005, 2009) calls on practitioners to think and act ecologically by focusing on ecological matters as a process, not a perspective added on to what they do. Dominelli (2002) emphasizes processes, power relations and contexts that include the physical environment as important dimensions of holistic practice that follows anti-oppressive principles including human rights and social justice. Social work practitioners and academics can become involved in tackling industrial pollution and environmental degradation in various ways.

Social workers' remit is not the technology behind these issues per se. However, it would help if practitioners understood the basic science behind disasters and the impact of these catastrophes upon populations and the physical, social, economic, cultural and spiritual environments that people live within, to support individuals and community groups seeking to redress or cope with these. Their activities include helping victim-survivors to: obtain immediate assistance like water, food, shelter and medical care; seek treatment for any ill-health that might result; rebuild devastated homes and communities; cope with the trauma brought about by being involved in industrial or physical environmental disasters; support endeavours to clean-up and decontaminate the physical environment; formulate plans to prevent such catastrophes from happening in future; develop action plans for when they do; and enhance individual and community resilience by preparing for future calamities.

They can assist residents in developing alternative development paradigms that integrate social and environmental justice in their action plans. Literature on what social workers do when practising ecologically according to the holistic approach used by green social workers is limited. The case study below explores their work in creating different approaches to understandings of their practice in degraded environments and its impact on emotional and environmental well-being.

Case study

Jason was a teenager who lived in a disadvantaged white British working-class community where residents were either on benefit or unemployed. Most families in the area had not experienced well-paid, permanent employment for several generations. The estate on which they were located was run-down, with violence on the street, burglaries and substance misuse being commonplace. Teenage pregnancy rates were among the highest in the country, despite government efforts to reduce these. The housing stock was poor; the built infrastructure, equally so. Public transportation was scarce and expensive. The community did not provide an urban environment that oozed confidence as an attractive place in which to live. Nor did it encourage internally driven innovation and change in its condition.

The inhabitants felt neglected by their government, alienated from society and isolated from its wider networks. Opportunities for them to change their realities were rare. However, they helped each other cope with their poverty by supporting each other whenever they could – in their families, peer groups and neighbourhoods. Few outsiders ventured into the community. They were seldom welcomed if they did, unless they proved their allegiance to the community and commitment to do something for it. Albert was one outsider who broke through this barrier.

Albert was a community worker in his early thirties when he rented a room from a lone mother in a house in a dilapidated street where the grass was uncut, gardens tangled and overgrown, and rubbish piled high in the yards. He was working on employment issues on another estate in a nearby town. Albert had been raised in a more affluent part of the country where his mother had loved her garden and was constantly caring for it, making it flourish and filling it with beautiful plants. Albert had inherited his love of the environment from her, and he felt despairing every time he walked the streets of his new home. One long holiday weekend, he decided he would clear up the yard and garden where he lived. For the whole of that weekend, he smiled at the mirth he created as he piled barrow-loads of rubbish in a skip and struggled to cut the waist-long grass with a hand mower that he had bought at a second-hand shop. Eventually, a neighbour felt sorry for him and offered him a scythe to help him do a better job.

As Albert had no idea of how to use this tool, the neighbour decided to help him, on condition that he did the same on his garden when he finished with this one. Albert readily agreed and the garden and yard were 'respectable' by the end of the weekend. He spent the next few months working on the garden in his spare time, planting flowers and shrubs and mowing the lawn as needed. His landlady showed no interest in what he was doing, although she was willing for him to continue with 'his project', as she called it. Albert kept his word and helped the neighbour to clear his garden too. When he bought too many plants for his landlady's garden, he passed these on to him. Albert saw this as 'good neighbourliness' and initially did not give it much thought. However, the transformation of these two gardens became the talking point of the estate.

Several other people came to ask Albert if he would work on their gardens too. Jason was one of these. Albert thought about this and decided that he could turn this into an opportunity to get people to transform their own gardens and acquire a sense of pride in their community's physical environment. So, he agreed on condition that each resident of every house that joined this project (except for babies, small children and older people who were frail) would become involved in the work. Five households agreed. Albert stuck to his word, and when one household failed to live up to this commitment, Albert ejected them from the scheme, leaving their garden clearance unfinished. It happened to be Jason's house. Jason felt very embarrassed by this turn of events, especially when he walked past the others that were now virtually all cleared. After a while, he went to see Albert and offered to bring two other households into the project if he helped him to finish clearing his own garden. Albert consented and so more houses now had gardens that the residents could enjoy. This they did, when they sat outside in the warm weather, and began to talk to each other while their children played in them. During these conversations, they began to talk about their estate and wondered whether they could get Albert to help them achieve more. Some felt that they could get their landlords to improve the housing stock and clean-up the environment more generally – especially as the estate had derelict houses and disused factories surrounded by contaminated land. Working on these could provide employment opportunities for their community. Albert had brought them a sprig of hope – to improve their landscape and create jobs within it. When Jason asked Albert to do so, he readily agreed to extend his work.

Albert's actions had energized some residents to enable place to become a space for enacting solidarity, and caring for, and being cared by, others. Their mobilization was a reciprocated involvement as each person had to do something to get help (unless this obligation had been waived for them). Albert had acted as a catalyst in this community, once he became accepted as belonging within it. He became an expression of that community and its aspirations. His attempt to improve the quality of their physical environment in a small way proved a 'success' and motivated people to think about other wider environmental issues that they could begin to tackle by enlisting his help. Having the hope to do something about one's circumstances is crucial to communities becoming confident and feeling empowered. In this sense, Albert was engaging in green social work without imposing his views about what the community needed. He acted as a role model who showed, by example, that although the estate had many problems, people could tackle some of them directly. That people become encouraged to think broadly and empower themselves is an important aspect of sustainable community development. Feeling empowered also assists people in releasing their resilience potential individually and collectively, and can turn mundane resilience into robust resilience ready to tackle structural inequalities.

This case study indicates how living in a degraded environment has psychological repercussions. It can sap people's motivation and prevent them from enjoying what they have around them. And it can trap them in a cycle of alienation and isolation that feeds into a community's sense of helplessness. Jason was in this predicament until he became determined to improve his situation. Albert's work is instructive in that it demonstrates how people do not have to engage directly in psychological counselling to address problems that have psychological implications. It also shows that the state of the physical environment has a direct bearing on how people feel.

Conclusions

Industrial pollution and environmental degradation will be detrimental to the development of any community. The loss of employment when firms leave, and the degradation caused by their failure to clean-up the environment when accidents occur, have long-term consequences for those living in affected communities. Communities, left to their own devices, cannot always cope with these effects. Mundane resilience has to become robust resilience and address structural inequalities and power and resource imbalances. The Bhopal example demonstrates this trajectory in the requirement for multinational companies to become accountable for the long-term impact of industrial accidents. However, robust resilience does not guarantee success in redistributing power and resources. The poor flows of information that follow nuclear disasters, whether companies remain *in situ* or leave, impede the formation of robust resilience and indicate that companies must become more transparent in and be held accountable for their dealings with government and the general public for recovery to progress.

Social workers can assist in these processes by calling for detailed information, greater accountability and transparency from firms and governments. They can support people when disasters occur and secure rights to compensation. And they can act as role models who can improve the environment in small ways that motivate communities to empower themselves and do more.

5

Climate Change, Renewable Energy and Solving Social Problems

Introduction

Climate change depicts a recent environmental crisis that embodies global interdependencies in which what one country does has significant outcomes for developments elsewhere. The benefits that the West has gained through industrialization processes have increased greenhouse gas emissions to the detriment of people living less industrialized lifestyles, particularly among poor people living in rural areas of the Global South. Environmental damage has resulted in claims for the 'polluters to pay' for cleaning up the environment and developing renewable green energy sources that will limit temperature increases below 2 °C. People without incomes or on low ones, wherever they may live, lack the money necessary for accessing the expensive technologies that would enable them to maintain or enhance their quality of life without external practical support, state intervention or business beneficence. This complex reality raises questions about what social workers can do to support poor people everywhere to enhance their quality of life and engage with the new green technologies, debates about climate change and the impact of these upon residents whose lack of financial resources excludes them from the renewable energy market-place.

In the UK, poor people's access to renewable energy and their capacity to transcend financial limitations to their doing so might occur as an outcome of the government's 'feed-in' tariffs, whereby the state pays individuals and companies to convert to renewable energy (Gubbins, 2010). The introduction of 'feed-in' tariffs carries the potential to alter access to this technology by enabling utilities companies to enter partnerships with poor individual householders and local communities to enable both parties to benefit without costing the Earth. The Big Society's emphasis on community interventions could facilitate such actions if its limitations could be overcome. Gubbins (2010) identifies several barriers in encouraging energy schemes under the Big Society concept. These are:

- changing the incentive framework currently in operation for renewable energy development from capital grants and revenue payments like the Feed-in Tariff (FiT) to the Renewable Heat Incentive announced in the Coalition government's Comprehensive Spending Review; and
- reducing or eliminating grant funding for voluntary groups.

Renewable energy groups had successfully campaigned for the FiT because it gave community groups and smaller firms an opportunity to provide renewable energy. For example, groups that could purchase generators that produced less than 5MW of energy could access the FiT scheme, but not the Renewable Obligation Certificate system because smaller sums of money were required. However, a problem with the FiT approach was that small groups or firms had to have access to capital to begin with. Capital funds are difficult for them to secure in the prevailing fiscal climate.

There are small demonstration projects that are tackling issues like fuel poverty and unemployment through the development of micro-renewable energy technologies. By doing this, they contribute to reductions in fossil fuel energy consumption and the development of self-sustaining energy communities. I consider such endeavours in this chapter because community social workers have been involved in these in both the Global North and the Global South, e.g., Gilesgate in the UK, Misa Rumi in Argentina (Dominelli, 2010a, 2011).

Social workers as a professional grouping have not played an active role in climate change policy debates initiated under the auspices of the UN, including talks linked to the Kyoto Protocols, until recently. The International Association of Schools of Social Work (IASSW), the International Council on Social Welfare (ICSW) and International Federation of Social Workers (IFSW) organized presentations during the Climate Change Conference in Copenhagen in December 2009 to reverse this situation. Their involvement in UN discussions in subsequent meetings initiated by the body called the Conference of the Parties (COP) continues. Social workers were present at meetings under COP 15 in Copenhagen, Denmark, in December 2009; COP 16 in Cancun in Mexico in December 2010; and COP 17 in Durban, South Africa, in December 2011. In this chapter, I explore these developments and argue that they provide the profession with the opportunity to expand the domains in which social workers are proactive and to help people who are experiencing increased flooding, droughts and other 'natural' disasters that are precipitated by climate change.

I also argue that social workers are obliged to understand the science behind climate change; engage in debates about the topic; work with multiple stakeholders; draw upon the knowledge developed by physical scientists; act as cultural interpreters passing on local knowledge to scientists, and scientific translators conveying scientific knowledge to local residents; and mobilize communities in initiatives that will turn them into energy-aware consumers who can bring new opportunities into their localities and contribute to transforming social and political priorities and policy-making regarding climate change.

Climate Change

Climate change is a contested term used to describe one of the key challenges facing contemporary societies. The concept highlights how the world's climate is changing as a result of greenhouse gas or carbon emissions caused by human industrialization and urbanization activities, which have released greenhouse gases as a result of burning fossil fuels. Greenhouse gases include water vapour, carbon dioxide (CO_2), methane (CH_4), nitrous oxide (N_2O) and chlorofluorocarbons (CFCs). By trapping infra-red radiation which occurs naturally in the Earth's atmosphere, these gases cause air temperatures to rise. Significantly elevated concentrations of these gases are produced through fossil fuel consumption, deforestation and agricultural and industrial processes. These contribute to changes in air temperature, precipitation patterns, ocean acidity, sea levels and glacial melting.

The Intergovernmental Panel on Climate Change (IPCC) estimates that natural processes account for only 5 per cent of climate change (IPCC, 2007). Measured in parts per million (ppm), greenhouse gases have risen from 280 ppm before the industrial revolution to 430 ppm by 2005, and are growing (IPCC, 2007). Each country will be impacted upon differentially as extreme weather events increase in frequency. These will produce substantial numbers of climate change refugees (Guzmán et al., 2009), both externally and internally to a particular country (Besthorn and Meyer, 2010). People in the poorest countries will face increased risk of flooding where weather gets wetter (and colder in some places) as sea levels rise; or a greater likelihood of drought where it becomes warmer and drier (UNDP, 2008, 2009). Both types of disasters will undermine livelihoods and cause extensive environmental damage, including loss of biodiversity and extinction of species, some of which will be irreversible. Thus, making industrialization processes carbon-neutral is essential for the health of people, the planet and its flora and fauna.

Demands for increasing standards of living throughout the globe, and reliance on industrial and economic processes that are driven by fossil fuels are expected to increase global demand for energy by 60 per cent between now and 2030 (Löscher, 2009). Meanwhile, IPCC scientists claim that the atmosphere is approaching its limits for absorbing emissions if rises in temperature are to be limited to 2°C or less. They have set this amount as a total of 1,400 billion tonnes of carbon emissions that the atmosphere can absorb between 2000 and 2050 (IPCC, 2007). Their calculations indicate that carbon emissions rose by 40 per cent during the last 200 years and predict that this will reach 550 ppm by 2035 if emissions are not reduced (Stern, 2006). To make matters worse, environmental stress will be intensified if methane currently contained in the permafrost of Siberia and Northern Canada is released in the rush to exploit oil reserves. Methane causes more atmospheric heating per unit than carbon dioxide (Löscher, 2009).

The Polluter–Victim Binary

The Kyoto Agreement has framed climate change discourses around the polluter–victim binary. Under it, the West is deemed responsible for precipitating the climate change crisis through its unbridled disregard of the consequences of fossil fuel-based industrialization. The remainder of the globe is cast as this polluter's victim. There is considerable merit in this point of view. As early as 1865, John Tyndall in the UK demonstrated that gases like water vapour and CO_2 retain heat. In Sweden, Svante Arrhenious warned that CO_2 emissions would lead to global warming as long ago as 1896. But neither policy-makers nor industrialists took their warnings seriously, and climate sceptics still do not (Giddens, 2009). Moreover, Western nation-states have failed to meet the greenhouse gas reductions stipulated in the Kyoto Protocol (World Bank, 2010).

The simple division between polluter and victim has become blurred as countries in the Global South embark on their own industrialization processes. Consequently, the balance in fossil fuel energy consumption is beginning to shift. This outcome occurs because industrialization processes in the Global South are also fuelled primarily by fossil fuels. This need not be the case if access to technologies based on renewable energy sources were to be made available cheaply. In 2005, the largest consumers of energy included industrialized and rapidly industrializing countries. The percentage of emissions from the top emitters was as follows: the USA, with 20.5 per cent; China with 15.0 per cent; Russia with 5.7 per cent; Indonesia with 4.7 per cent; Japan with 3.0 per cent; Germany with 2.4 per cent; France with 2.4 per cent; Canada with 2.4 per cent; the UK with 2.0 per cent; South Korea with 1.9 per cent. China overtook the USA as the major emitter in 2006 and the two of them are now responsible for 40 per cent of overall emissions in the world. In the year 2008, the emerging economic giant of China emitted 6.1 billion tonnes of CO_2. Its aim of removing 250 million people from poverty will mean that China's emissions will rise to 10 billion tonnes by 2020 (Löscher, 2009: 29). The issue of emissions is further complicated by the degree of efficiency in the use of units of energy to produce commodities. For example, China uses 3.5 times more energy than the global average for each unit of GDP produced (Löscher, 2009).

Rapid population growth will intensify pressures on all resources to meet ever-growing needs and thereby impact upon climate change. Population growth is highest in the Global South (UNDP, 2009). The number of people in China will rise even more if the government ends its one-child policy (even though observed primarily among its urban population) to redress a gender imbalance between the numbers of men and women (Guzmán et al., 2009). Similar issues arise for the populace of India where boy babies also outnumber girl babies. Increases in the size of the middle class in both India and China, and their goal of adopting Western lifestyles will place further pressure on the Earth's resources. Their demands for improved standards of living are reasonable, but cannot be met using an industrial model that Western countries developed for substantially smaller populations. Squaring the circle between requirements for fuel and its supply requires considerable investment in renewable sources of energy that do not exacerbate conditions that produce climate change. Current fuel usage is pulling in the opposite direction, especially among emerging economies because their adoption of Western middle-class lifestyles among much larger population sizes will eventually result in their overtaking the West in energy consumption on both per capita and national bases.

India and China together hold one-third of the Earth's population within their borders. The development of more sustainable, environmentally friendly forms of socio-economic development is a must for these two countries and the rest of the world if equality and decent standards of living are to be made available to all of the Earth's inhabitants regardless of locale, nationality or other social division. Addressing these complexities requires cooperation among the world's nations to find solutions to the pressing socio-economic, political and environmental problems that will affect us all. Social workers have a role to play in: facilitating an understanding of these issues; convincing people of the importance of focusing on environmentally embedded, eco-friendly forms of socio-economic development; working with local people to develop sustainable solutions to social problems; encouraging the sharing of green energy technologies; and lobbying for policy changes that will transform the current socio-industrial political order.

People's responses to climate change

Public discourses on climate change divide people into a binary of those supporting the idea of a people-generated impact on the world's climate (the greens) and those rejecting it (the sceptics) (Giddens, 2009). This is a simplistic categorization of the range of opinions that prevail. In the UK, a survey by DEFRA (the Department for the Environment, Food and Rural Affairs) revealed a wider spectrum of views among the British public and clustered people's responses as follows:

- *Positive greens* – 18 per cent of respondents who would do their utmost to limit their impact on the environment;
- *Waste-watchers* – 12 per cent of respondents whose thrifty lifestyle meant they recycled extensively;
- *Concerned consumers* – 14 per cent of those replying who believed they were already doing a lot and would not do more;
- *Sideline supporters* – 14 per cent of those surveyed who acknowledged the problem of climate change but insisted on maintaining current lifestyles;
- *Stalled starters* – a group with little information about climate change, who could not afford an affluent lifestyle but aspired to it;
- *Honestly disengaged* – these respondents were disinterested about the matter and thought it was irrelevant to their lives.

Only 23 per cent of Britons prioritized climate change as the most important global social issue, while a further 58 per cent considered it one of several serious matters to be tackled (Giddens, 2009).

Climate Change Endeavours Led by the United Nations

The UN, as the organization that brings nation-states together to deliberate on international issues in one world body, has played an important role in facilitating discussions on climate change. The UN created the United Nations Framework Convention on Climate Change (UNFCCC) in 1994 as an international treaty to get countries to lower greenhouse gas emissions. Its endeavours focus on international negotiations among governments. These are conducted primarily through what is known as the COP. Meanwhile national governments ensure that firms and individuals within their borders comply with the reduction targets set. Moreover, these governments agree to support access to green technologies for all countries. The COP met first in 1995 and has done so annually since then to carry this agenda forward. There were 189 members of COP by 2006. The 2011 meeting or COP 17 was held in Durban, South Africa, and aimed to take forward the agenda agreed at COP 16 held in Cancun, Mexico, in 2010. A worry among small nations in the Global South and civil society organizations throughout the world since Copenhagen, where COP 15 was held in 2009, is that aspirations for a legally binding agreement remain beyond reach.

Meanwhile, the UNFCCC had already established the Subsidiary Body for Scientific and Technological Advice (SBSTA) and the Subsidiary Body for Implementation (SBI), a secretariat and financial structures to enable COP members to carry out their work. The structure of the SBSTA was enhanced in 2000. In 2001, the Marrakech Accord affirmed this development and gave adaptation higher prominence. The UNFCCC architecture was restructured again in 2004 when COP 10 established two tracks for adaptation measures. One of these focused on a five-year programme of work that examined the scientific, technical and socio-economic dimensions of vulnerability and adaptation to climate change. The other programme considered adaptation activities, technology transfers and capacity building. This development became enshrined in what became known as the Buenos Aires Programme of Work on Adaptation and Response Measures, and marked a significant shift away from an exclusive concern with reducing carbon emissions. This development was followed with recognition of the needs of the 48 Least Developed Countries (LDCs) and their being charged with formulating their National Adaptation Programme of Action (NAPA) to access adaptation funds. Oxfam estimates that they will require US$50 million per year to adapt successfully. In 2006, the Buenos Aires Programme was renamed the Nairobi Work Programme on Impacts, Vulnerability and Adaption to Climate Change. The lack of legally binding emissions targets and a timetable for action among all countries were painfully evident during the third COP meeting in Kyoto, Japan, although the Kyoto Protocol was agreed there and has continued to dominate COP discussions since that time, without a further legally binding accord being reached.

Kyoto and beyond

The foundation stone of the UN's initiatives on emissions reductions is the Kyoto Protocol, which was signed by 184 countries in Kyoto, Japan, in 1997. Kyoto came into force in 2005. It included a fund for adaptation aimed at mitigating the impact of climate change, and required Annex 1 countries, or 37 of the richest industrialized nations, to reduce carbon emissions to 5 per cent below 1990 levels between 2008 and 2012. The ideas behind Kyoto initially emerged in 1992 with the Earth Summit in Rio de Janeiro, Brazil, where participating governments formed the UN Framework Convention on Climate Change (UNFCCC) which came into force in 1994. Under the Rio Accord, governments agreed to limit rises in the Earth's temperature to less than 2°C. The West's acceptance of culpability in initiating climate change underpinned this Agreement. This became known as the 'historical debt' which Annex 1 countries were committed to redressing. As part of Kyoto, the rich industrialized countries became committed to reducing greenhouse gas emissions and helping industrializing countries financially and through technology transfers.

Implementing the Kyoto Agreement became problematic from the beginning. Some rich countries failed to ratify the Kyoto Protocol. The American Senate rejected it and George W. Bush withdrew the USA from the Protocol in 2001. Australia's Parliament did likewise that year. Moreover, some countries that ratified have failed to meet their commitments. For example, Canada, committed to a 6 per cent reduction in emissions from 1990 levels, had these rise by 25 per cent to develop Alberta's oilsands.

Under Kyoto, emerging economies had no responsibility for keeping their emissions low when pursuing an industrialization strategy to raise standards of living for poor people. Industrializing countries were to reduce emissions voluntarily by participating in the Clean Development Mechanism (CDM), funded through an Adaptation Fund consisting of a 2 per cent charge on CDM projects. The absence of requirements for emerging economies and nations aspiring to industrialize in the future either to control their emissions in the course of industrialization or to utilize renewable energy technologies in order to achieve their objectives created a polluter–victim binary within Kyoto. Reaching a consensus on either of these possible ways forward would have been difficult because the industrialized nations that were, at that point, responsible for most of the greenhouse gas emissions under discussion would have had to agree to pay either to clean up the environment quickly to secure Kyoto targets or to transfer green renewable energy technologies to the rest of the world so that it could industrialize without the huge rise in carbon emission endemic to a path of industrialization based on fossil fuels.

Casting industrialized countries as 'polluters' and industrializing countries as 'victims' created tensions between those who had to reduce emissions and those who did not, and pre-empted the pursuit of a more inclusive and forward-looking policy that acknowledges that, as the world is a shared one, there are obligations for all countries to help each other to achieve a decent standard of living for all the planet's inhabitants without costing the Earth. While some industrializing countries became actively willing to shoulder the burden of paying off this historical debt, others were passively engaged and neglected to take action that would reduce their emissions. These tensions have proved difficult to shift in subsequent meetings.

A recent report by McKinsey Consultants has criticized the UN for poorly administering the CDM and not monitoring adherence to the Kyoto Protocol and its timelines. Consequently, emissions rose substantially in countries such as China, India, Brazil, Mexico and South Korea, as well as in several industrialized countries that had ratified Kyoto.

Reaching agreement on the monitoring mechanisms also proved elusive. The Marrakech Agreement on the methodology for monitoring Kyoto was not concluded until 2001. The National Adaptation Programme of Action (NAPA) for each country formed the central plank of the Marrakech Accord. NAPAs can be used to assess and address the impact of increased population size and climate change on: food security/insecurity; depletion of natural resources; environmental degradation; water scarcity; human ill-health; migratory movements; and urbanization of rural areas within individual countries. The formulation of NAPAs provides opportunities for social worker involvement because governments are required to consult with local communities on the actions that have to be implemented to reduce greenhouse gas emissions, and social workers can assist in achieving this goal. NAPAs also emphasize adaptation as a key strategy in addressing climate change. According to the IPCC (2007), adaptation is an 'adjustment in natural or human systems in response to actual or expected climatic stimuli or their effects, which moderates harm or exploits beneficial opportunities'.

It took until the Bali meeting in 2003 to formulate and agree the timetable for a successor to the Kyoto Protocol, which expires in 2012. At the Poznan Climate Conference in 2008, progress in resolving the impasse around the successor to Kyoto was virtually halted as politicians awaited the result of the American Presidential elections. When Barack Obama won, hopes rose that he would support tackling climate change and agree a new Protocol at the Copenhagen Summit on Climate Change in December 2009. Unfortunately, no binding agreement materialized in Copenhagen. Although Obama secured China's agreement to reduce its own emissions (Averchenkova, 2010), the 'Copenhagen Accord' he brokered did not constitute a legally binding agreement. Nor did it presage collective action among nations. Instead, it encouraged each country to set its own limits and have its plans available for discussion at COP 16 in Cancun, Mexico, in 2010 (Cryderman, 2009). However, no legally binding international agreement was forthcoming in Cancun or Durban subsequently.

Other compliance mechanisms proved problematic. Carbon 'credits', developed in the USA, were to incentivize private firms to reduce emissions and become less polluting by setting the industry's price-tag for pollution. This proposal failed to achieve these aims because entrepreneurs can earn millions of dollars by selling 'credits' in a carbon trading scheme (CTS) whereby a polluter buys a non-polluter's carbon units in exchange for their polluting ones. By enabling firms to purchase 'carbon credits' held by non-polluting companies, CTSs create a market whereby polluting industries can continue polluting by 'selling' their pollution (Friends of the Earth, 2008). CTSs are ineffective because they reward polluters; enable certain groups to profit from selling carbon credits without reducing overall emissions; ignore those who pay the price when nothing is done, e.g., small island nations sinking into the sea; and allow fraudsters to profit from their operation. The European Emissions Trading Scheme (ETS) erroneously rewarded Eastern European polluters that sold carbon credits to Western companies that then lacked incentives to lower their emissions. Fraudsters targeted the ETS by claiming and reclaiming VAT, thereby obtaining about 90 billion euros per year. And, in 2011, a cyber attack resulting in the theft of £28 million in carbon allowances forced the closure of the ETS (Macalister and Webb, 2011).

Private firms could amass further financial gains, without doing much to reduce pollution levels, through public subsidies. This was instanced by the provincial government in British Columbia in Canada which spent CDN$14 million in 'seed' money to create the Pacific Carbon Trust (PCT), and a further CDN$869,000 to offset the anticipated 34,370 tonnes of carbon emissions it would produce. The scheme subsidizes firms using clean, green technologies by paying an undisclosed amount per tonne of carbon reduced. The scheme charges public-sector agencies, especially schools and hospitals, CDN$25 per tonne of carbon emitted as a 'fine' to fund the scheme, although monies available to enable public-sector bodies to purchase less polluting technologies are limited. The 'fines' collected go to private firms and earn them profits (Bader, 2009). According to Bader, public enterprises in British Columbia could save tax dollars by using the Chicago Climate Exchange Scheme which charges only US$0.14 per tonne to off-set carbon emissions. This is considerably cheaper than the government's scheme! Although 14 cents is a paltry incentive to reduce emissions, it becomes substantial if calculated across huge tonnage. Despite these failures in market-oriented initiatives, UN Executive Secretary Yvo de Boer claimed during COP 15 in Copenhagen that the market was more efficient than the state in reducing carbon emissions, and supported market-based mechanisms over taxes and regulation for doing so. Large polluting private companies like Exxon Mobil share this view.

Agreeing the costs of and paying for cutting emissions globally is another contentious area. The 2006 Stern Report informed rich countries that tackling climate change then would cost less than 1 per cent of world GDP. This figure would rise to 20 per cent of GDP if significant measures were left to a later date. Facing the lack of movement by the rich countries, Lamumba Stanislaus Di-Aping, of the Sudan and chief negotiator for the G77 (group of poor industrializing countries and China) that encompassed 132 of the 192 countries attending COP 15, argued that the International Monetary Fund (IMF) and World Bank should *not* run the proposed 'climate fund' and that lack of a deal at COP 15 in Copenhagen meant 'certain death' for Africa. Di-Aping thought derisory Gordon Brown's proposed annual budget of US$10 billion to fund climate change adjustments in industrializing countries and added that it would not buy 'poor nations the coffins' needed if greenhouse emissions were not lowered (*Daily Telegraph*, 9 Dec 2009: 14).

To progress reductions in carbon emissions, the EU proposed that 100 billion euros were needed annually until 2020. It suggested dividing payments according to calculations based on the size of GDP and linked to the level of carbon emissions: US$30 billion from Europe; US$25 billion from the USA; and the remainder from the rest of the industrialized world (*The Week*, 7–13 December 2009: 28). The EU determined that this allocation was affordable because it comprised less than 0.3 per cent of the annual overall income of rich countries. Extensive dissent over this proposal caused Danish Prime Minister, Lars Løkke, who hosted COP 15, to propose 'one agreement, two steps'. Under his suggestion, COP 15 negotiators would agree the outline of a Treaty in Copenhagen, but finalize details at COP 16 in Mexico. This strategy nearly achieved a consensus after two weeks of heated deliberations in which those most affected by the lack of progress argued strongly for a legally binding agreement. It was squashed when the USA, China, India, Brazil and

South Africa brokered the Copenhagen Accord. Small island nations facing annihilation by being submerged by rises in sea levels refused to sign it, e.g., Tuvalu. Resistance was high, as ninety countries refused to sign this Accord and were later threatened with losing Adaptation Funds as a result, e.g., the US threatened to withdraw US$3million of its intended aid from Bolivia and US$2 million from Ecuador (Vidal, 2010). And although South Africa helped to formulate the Copenhagen Accord, the opposition at home dubbed it the 'Hopelesshagen Flop'. The 'greens' criticized the Copenhagen discussions for not realizing the ambitions of Kyoto. Carbon emissions globally are currently 25 per cent higher than in 1990. Annex 1 countries are now responsible for 25 per cent of global emissions, a figure that overestimates progress by omitting the USA, a major polluter that withdrew from the list in 2001.

These responses indicate that the political will to reduce greenhouse gas emissions substantially and to share technological developments is absent. Social workers, skilled in holistic interventions, could facilitate implementation discussions by mediating between conflicting groups in local communities and at policy levels nationally and internationally. The IASSW, ICSW and IFSW hold consultative status at the UN and could suggest alternative policies in that body to help negotiations move beyond the impasse reached at COP 15 (Averchenkova, 2010). They are planning to hold discussions on issues including climate change at their World Congress in Stockholm, Sweden, in 2012, when they deliberate upon their Global Agenda. This Global Agenda will also be presented to the UN during Social Work Day in 2012.

Despite the overall gloomy picture, Copenhagen 2009 revealed that politicians and environmentalists agree on the nature of the problem and importance of physical limits to on-going pollution. However, they disagree strongly on how to curtail it. Politicians favour market-based solutions associated with carbon trading schemes (CTSs) rather than state regulatory ones. Yet, markets have proven unreliable in this regard, and civil society organizations do not trust the market to endorse action that will reduce temperature rises to no more than 2°C between now and 2050.

Hope of success glimmered briefly in December 2009 when the Environmental Protection Agency in the USA ruled that carbon dioxide is a health hazard. Thus, Senate, which vetoed the Kyoto Protocol, does not have to approve any carbon trading scheme that President Obama signs (Mason, 2009). Yet, there has been little progress in getting American commitment to a legally binding agreement, despite this development, during the intervening years. However, some local responses have provided optimistic notes as local communities aim to achieve what national rulers cannot. For example, British residents formed the 10–10 Campaign to reduce emissions by 10 per cent during 2010. Anyone sharing this objective could join: individuals, companies and local authorities. Over 100 local authorities had subscribed to it soon after it was set up. Social workers can use local situations to help people understand wider global concerns and mount consciousness-raising campaigns that encourage local action to reduce fossil fuel consumption, as occurs in the following narrative.

Case study

Deprived communities in the UK have high levels of fuel inequalities. A small project in Gilesgate in England sought to tackle this issue by developing renewable sources of energy in the community in the hopes of creating cheaper sources of energy and employment opportunities for local residents. A community social worker assisted the community to reach its goals by bringing together a partnership of local residents, an interdisciplinary team of academics and students from the local university, practitioners from a range of professions including social work, policy-makers, housing providers and private businesses investing in micro-generated renewable energy sources. They focused their attention on reducing fuel poverty and fuel inequalities, facilitating access to renewable energy sources and cutting their carbon footprint, developing sustainable employment opportunities, and co-producing knowledge through collective action.

The partnership held public meetings and exhibitions where residents examined renewable technologies and considered their options in energy production and consumption. They had energy audits of their homes and learnt what grants and subsidies the government funded to incentivize communities in becoming energy-aware and engaging with the latest developments in green energy, and reducing and monitoring energy consumption in their homes and community edifices.

Some residents began to adapt their homes and and encouraged those running public buildings to lower energy consumption to reduce fuel bills. And they sought to get private firms to commit to manufacturing clean technology and renewable energy products in their locality, to create jobs and save energy. Their long-term aim is to become energy self-sufficient, feed into the national grid, manufacture green goods for the wider society and reduce their carbon footprint.

The community social worker was central in developing capacity and social capital – bonding, bridging and linking, by coordinating the different stakeholders, organizing meetings and exhibitions, and facilitating the collection of data for monitoring purposes from different players. These became the basis of the connections that enabled people to realize their common objective of 'making a difference' by enhancing community well-being and looking after the environment for themselves, their children and their grandchildren.

This project reveals that people can seek equitable solutions that draw upon green technologies to improve the quality of their lives and limit the amount of greenhouse gases that enter the air, waters and soils of the planet. Clean technologies create jobs, thereby alleviating poverty and contributing to the realization of human rights, and social and environmental justice. In 2010 at the climate change discussion in Bonn, Bolivia argued for this approach and included social workers among those advocating for them (TWN, 2010).

Environmental Justice

Industrial pollution raises serious questions about environmental justice. What it means and who benefits from it are particularly contentious points. Bullard (1990) highlighted environmental racism, or the processes whereby black and other disenfranchised minority communities bore a disproportionate share of the effects of environmental pollution, when he exposed the raced and classed basis of the dumping of industrial effluent caused by the disposal of toxic industrial waste, primarily in poor communities. Others have followed in his footprints and enlarged his ideas. For example, environmentalists in the Global South have exposed the damage done by mining and other extractive industries, especially in pristine forests, jungles and rainforests (Liebenthal, 2005). Although these activities often promised jobs for local residents, these have proved elusive, and such initiatives have often lined the pockets of rich and corrupt politicians (Denton, 1986).

Anthropomorphic definitions of environmental justice abound. The Environmental Protection Agency (EPA) in the USA has defined environmental justice as:

> the fair treatment and meaningful involvement of all people regardless of race, color, national origin, or income with respect to the development, implementation, and enforcement of environmental laws, regulations, and policies. EPA has this goal for all communities and persons across this Nation. It will be achieved when everyone enjoys the same degree of protection from environmental and health hazards and equal access to the decision-making process to have a healthy environment in which to live, learn, and work. (www.epa.gov/environmentaljustice)

While its egalitarian thrust is welcome, it fails to recognize the significance of caring for the environment at the same time as addressing structural inequalities. Writing for environmental activists, David Schlosberg (2007) contends that environmental justice encompasses four major tenets. These are:

- distributing environmental risks and benefits fairly and equitably;
- having local residents participate fully in decisions about their environment;
- acknowledging local community cultural traditions, knowledge and ways of life; and
- recognizing the ability of communities and individuals to make their own choices effectively and succeed within their societies.

These principles are invaluable in the interpersonal domain. Their scope is limited because they ignore the 'structural' inequalities engendered by the current mode of industrial production. Although environmentalists focus on the rights of ethnic minorities and lack of voice for poor and marginalized populations, these marginalized groups have criticized the environmental movement for being:

- composed primarily of middle-class activists and excluding marginalized populations, including indigenous peoples, whose views for sustaining the environment can vary substantially from theirs;
- ideologically conservative and favouring the status quo; and
- socially regressive in its socio-economic and political impact (Escobar, 1998; Hodgson, 2002).

These insights have provided the foundation stones for environmental activists in several countries. Environmental stances taken by activists seeking to restrain middle-class-dominated activism demonstrate agency among poor people and highlight how they are involved in making choices about their habitats that are consistent with their own worldviews and what they want to get out of life. They are not passive victims of other people's interventions, and they do not necessarily prioritize capitalist methods of production in meeting their daily needs, e.g., indigenous peoples in the Amazon Basin (Escobar, 1998).

The environmental movement includes social workers, and particularly those doing community development work. Increasingly, the environmental movement is acting globally because multinational corporations know no borders, as they move their plants and jobs around the world to extract the maximum subsidies and employ a poorly unionized or non-unionized workforce in a few local jobs that are badly paid. Additionally, environmentalists cross borders to ensure that solidarity in environmental actions pursues egalitarian goals and collective action for the benefit of people and the environment all over the world (Escobar,

1998). Environmental NGOs (non-governmental organizations) are an important part of the environmental movement and support poor communities in accessing renewable green technologies to reduce greenhouse gas emissions, protect the environment and enhance the quality of life of poor people, as the case study below indicates.

Case study

In Misa Rumi, Argentina, an indigenous llama-herding community installed solar-fired stoves in their homes to replace those using firewood to cook and to heat mud-brick homes. The practice of collecting firewood had led to deforestation and soil erosion. Desisting from these practices could protect the yareta tree that takes hundreds of years to grow. The residents achieved this goal under the auspices of a local NGO, the EcoAndina Foundation. Its credentials had been established in a close working relationship that stretched back to 1989. Working together made it possible for residents to use solar power for domestic purposes, civic buildings and irrigation. This has enabled the community to reduce carbon emissions and earn carbon credits for having reduced greenhouse gas emissions. Each single solar-powered cooker saves 2 tonnes of carbon dioxide a year and their use by 40,000 people yields considerable savings in pollution caused by burning wood (Stott, 2009).

The people in Misa Rumi have demonstrated that social and environmental justice go hand-in-hand. By acting in consciousness-raising or educational roles, social workers can become catalysts that make local people aware of the green technologies that they can access to solve their problems, or link them up with engineers that can develop the technologies that they need. Lane et al. (2011) discuss co-producing solutions by crossing disciplinary divides and engaging experts in dialogues with local communities to promote innovative responses to flooding, an extreme weather event linked to climate change, for example (Lane, 2008).

Social Work Action on Climate Change

Social workers are involved in action on climate change at individual, community and international levels. They can raise awareness about climate change and find solutions to personal problems like experiencing fuel poverty through renewable energy initiatives that bring people and resources together in specific actions, including the collective ones illustrated by the Gilesgate Project (Dominelli, 2011). Personal action alone is not enough in solving structural inequalities including fuel poverty or climate change. Collective solutions, based on developing consensus among participants at the local, national and international levels, can help resolve global problems locally and overcome blocks to reaching agreement over a legally binding treaty globally. The *Equitable Carbon Sharing Scheme* (ECSS) that social workers approved in Copenhagen on 9 December 2009 aims to transcend the stalemate caused by the blame game about who is responsible for causing the climate change problem and solving it by focusing on interdependencies between peoples and their environments (social, physical and natural), addressing issues at the micro-, meso- and macro-levels, and including all individuals equitably within a human rights and social justice framework that acknowledges peoples' obligations to their physical environment, other sentient beings and themselves.

The Equitable Carbon Sharing Scheme: an international response

The Third World Network's (TWN's) daily summaries of discussions at COP 15, COP 16 and COP 17 revealed that the 'rich' country (polluter) – 'poor' country (victim) binary flounders because no country will risk reducing its own emissions without the others doing likewise simultaneously. The ECSS reduces this risk by: building a consensus based on global interdependencies and responsibility; acknowledging the finite amount of carbon emissions that the Earth can absorb if temperatures are not to rise by more than 2°C, which has been calculated to be 1,400 billion tonnes by 2050 (Stern, 2006); sharing the Earth's resources and technical know-how equitably; including the world's current and future populations in the equation; recognizing people's duty to care for the environment alongside the right to be cared by it; and holding firms accountable for the decisions they make concerning the development, or more often the exploitation, of the Earth's resources. This duty to care for Planet Earth and be cared by it through its supply of material and physical resources extends the notion of environmental justice beyond that associated with ensuring that marginalized groups obtain their fair share of resources. Additionally, it considers the Earth's beneficence as a public, not private, good, and, therefore, accessible for meeting the human needs of all as defined by the UDHR, rather than swelling the bank accounts of a monied few.

The Earth's population is anticipated to grow beyond 9 billion by 2050 (UNDP, 2009), necessitating both individual and collective action if the sharing of finite resources equitably among current and future generations is to materialize. Distributing these goods equitably intergenerationally requires: that pollution associated with each individual's carbon emissions encompasses all needs including manufacturing processes, transportation, housing, heating, lighting, growing food, and provision of services like health, social care, education and defence, with each person having the same allocation of greenhouse gas emissions regardless of status or residence; the *curbing of pollution* in current consumerist approaches to meeting human need perpetrated by both industrialized and industrializing countries; and the rapid sharing and deployment of renewable technologies. Reaching parity requires those currently consuming more than their share of energy to reduce their usage, whilst allowing those who do not to increase theirs. Mathematical models can assist in the task of identifying and forecasting consumption for each individual. Social workers can liaise with mathematicians to bring these types of data into the public domain and translate these models into language that people can understand and use (Dominelli, 2011).

Implementing ECSS

Implementing ECSS on an equitable basis requires that rich people everywhere (whether in the Global North or South) and high fossil fuel energy consumers in the Global North and Global South *reduce* substantially their carbon emissions. Poor people in the Global South with low current energy consumption would be able to *increase* theirs and draw on renewable energy sources to create a sustainable development that would enable them to rise out of poverty. This approach takes account of the historical privileging of the West's energy use, energy consumption by the emerging economies, and the need to raise the living standards of the world's poorest people now and eradicate income and fuel inequalities for future generations.

Money being misspent on polluting the Earth could be utilized to foster clean technologies by running down polluting ones and declaring a construction moratorium on environmentally damaging forms of energy production and consumption. ECSS advocates for the free transfer of clean renewable technologies to enable each person to secure energy in environmentally friendly ways. Entrepreneurs would be able to run profitable companies even if they gave free access to these technologies and charged only for the end product because a large part of their research and development costs have already been subsidized through the public purse because governments use tax monies to provide transportation networks, other infrastructures and grants to develop renewable technologies.

Risk Reduction

Risk reduction strategies have become concerned with how better to anticipate, reduce and manage disasters by linking risk reduction with sustainable development rather than focusing primarily on humanitarian action after disasters have occurred. This requires knowledge of different hazards, of how to strengthen the coping and adaptive capacities of different vulnerable populations, and of systematic approaches to mitigating risk. To build robust resilience in these circumstances, social workers need strategies that deal with prevention, mitigation, preparedness, response, rehabilitation and recovery for both current and anticipated hazards. They also have to work with experts from other disciplines to ensure that they have all the knowledge that they need at their disposal and become capable of translating scientific jargon into accessible language for community groups.

According to the UN (2006), the shift from disaster responses to risk reduction has been prompted by the rising costs of disasters in terms of the damage caused to people, and socio-economic, environmental and biological systems. To raise public consciousness of disasters, the UN declared 1990–9 the International Decade for Natural Disaster Reduction (IDNDR).

The first World Conference on Natural Disaster Reduction took place in Yokohama, Japan, in 1994. The Yokohama Strategy and Place of Action for a Safer World urged individual countries to develop the infrastructures necessary to mitigate the impact of disasters. The UN General Assembly replaced the IDNDR in 2000 with the ISDR. Involving a coalition of governments, UN agencies, regional organizations and civil society organizations, the ISDR sought to develop and sustain strong international action in reducing the risks that accompanied disasters.

The second World Conference on Disaster Reduction took place in Kobe, Japan, in 2005. It widened its focus, shifted the emphasis to preventative action and introduced the *Hyogo Framework for Action 2005–2015: Building the Resilience of Nations and Communities to Disasters*. The Hyogo Framework aims proactively to inform, motivate and engage people in disaster risk reduction in their local communities, and to involve the international community in addressing the issues of:

- governance including organizational, legal and policy frameworks;
- risk identification, assessment, monitoring and early warning;
- knowledge management and education;
- reducing underlying risk factors; and
- preparedness for effective response and recovery (ISDR, 2005).

The Hyogo Framework suggests that these concerns are integrated into all considerations of sustainable development, environmental protection, and disaster planning and management. Moreover, the ISDR system is being strengthened by the Global Platform for Disaster Risk Reduction which has a dedicated fund to support its implementation.

The International Association for Impact Assessment (IAIA) has defined the social impact of climate change as 'all impacts on humans and on all the ways in which people and communities interact with their socio-cultural, economic and biophysical surroundings'. As climate change is experienced differentially according to context, geographic features and population, assessing it for specific groups in particular places requires sensitive and detailed calculation (Spickett et al., 2008). Measuring these impacts reveals that social vulnerability is increasing, especially in creating a growing demand for health and social care services to overcome poor health outcomes. Those with mental ill-health and/or disabilities may find that provisions become inadequate for their needs unless governments and other providers begin to plan for increases in these areas now (CAG Consultants, 2009). Providers also have to prepare for rising psychological stress among those who endure the consequences of climate change, manifest as extreme weather events such as flooding, heat waves and cold spells (Reacher et al., 2004). The situation can be exacerbated if the health and social care infrastructures are deleteriously affected simultaneously. Older people, especially those located in coastal areas where significant proportions of them live, are particularly vulnerable (McMichael, 2006), and so are children (Mitchell et al., 2009).

Having holistic approaches to climate change and involving the wider community in people's deliberations is crucial in supporting human rights and social justice-based interventions that value and respect the environment. A good example of this approach comes from the Scottish Isle of Eigg. In this illustration, people became involved in collective action that considered issues of sustainable development that encompass current and future needs of the population and the environment. Community ownership, control and management of assets, including land, are central to such endeavours (Aiken et al., 2008). The Eigg initiative is described below.

Case study

The Isle of Eigg Heritage Trust was created as a charitable organization in 1996 to encourage community participation in decision-making about how to transform land ownership and determine the future development of the Island. It involved the community in a buy-out of private land-owners in

1997, to end their exclusion from decisions about the socio-economic development of Eigg. Deliberations about future developments were to include the diverse range of opinions and different skills available on the island and to ensure that there was a balance between development and securing the financial stability needed to maintain everyone on Eigg and conserving its wildlife and beautiful physical environment.

Initial difficulties in communication, participation and attendance at meetings were thrashed out and a consensus reached that catered for the more informal ways that Islanders preferred to conduct their meetings. At the same time, a record of decisions taken was made to support the directors and workers of the Eigg Heritage Trust. Additional meetings and workshops involving the community were also called as needed. This more participative approach enabled the Trust to reach agreement on the leases to the land held by individuals, and to construct the Pier building which included a tea room built to their design and liking. They also constructed a 6Kw hydro turbine to feed into the Eigg electricity grid and reduce reliance on fossil fuels. Local contractors and workers were used wherever possible.

The Trust received voluntary help to deal with financial matters, planning and fund raising. However, having a paid worker who could take these issues on board would have expedited matters considerably (Conway, 2010).

Eigg's example highlights inclusive robust resilience responses to social problems. Social workers become involved in such initiatives as residents and workers supporting culturally relevant collective action that protects the physical environment and looks after people's needs.

Conclusions

Climate change poses some of the most challenging issues that have to be addressed in the twenty-first century. Social workers can encourage understanding of the issues, translate scientific knowledge and jargon into accessible language and, as community development workers, to facilitate the achievement of robust resilience and promote community innovation and ingenuity. The Isle of Eigg Heritage Trust is a good example of how community initiatives can be inclusive, environmentally friendly and true to local cultures and traditions, while respecting the physical and natural environment and providing jobs that are sustainable.

6

Environmental Crises, Social Conflict and Mass Migrations

Introduction

Climate change, industrial processes that encroach into non-commercialized farmlands and forest lands, and agribusiness have exacerbated the loss of traditional habitats and lifestyles, particularly locality-specific nomadic or aboriginal ones that eschew consumerism and degradation of the physical environment. Such damage has led to a decline in traditional grazing lands, desertification and a range of other environmental crises that have resulted in conflicts, including violent clashes between groups of people. These changes have often resulted in mass migrations that have intensified stress on land in rural areas and built environments in cities as different populations come into close proximity with each other and compete for scarce social and physical resources.

In this chapter, I consider how social workers have become involved in issues that link environmental degradation, social tensions or conflict, and mass migration. These practitioners, often involved as mediators, development workers and therapists, assist people in rebuilding lives and communities after disasters in more sustainable and environmentally friendly ways. An interesting example that I explore is that of social workers intervening in the Mathare Valley in Kenya where the IASSW

(International Association of Schools of Social Work), ICSW (International Council on Social Welfare) and IFSW (International Federation of Social Workers) worked together with the UN HABITAT to bring harmony into an urban area in which conflict arose when nomadic peoples who had lost their grazing lands to drought and desertification settled in it. These nomads had migrated to the slums of Nairobi where conflict between the nomadic newcomers and more established urban dwellers ensued. The IASSW, IFSW and ICSW secured some funding from HABITAT, and, working in partnership with this organization, enabled local social work practitioners and academics to join community development workers in the slums and promote understanding between the groups involved in the dispute.

The Impact of Environmental Crises on People's Movements

Industrialization has produced environmental degradation and social costs that people, mainly poor people, have had to bear. Many of these costs have been invisible, others downplayed by economic balance sheets that fail to include the impact of industrialization and loss of environmental amenities on people, livelihoods, homes, land and ancestral belonging associated with a sense of place. Part of this invisibility is due to the exclusion of certain costs from economic models and their quantitative measures of the impact of environmental change. Subjective experiences of loss and the costs of trauma borne by individuals, plants and animals, the significance of ties to place and networks of social capital are among these. They are captured more readily qualitatively than quantitatively. Community activists have raised awareness of these issues and broadened discourses about what should be covered in assessing environmental damage and demanded that these calculations become sensitive to the social impact of economic decisions and their full environmental implications in both the long and short terms. For example, Arundhati Roy has advocated against the building of a gigantic dam on the Naramda river in India because it would destroy local peoples' lives and beautiful physical environments around the Sarda Sarovar Reservoir. Vandana Shiva (2003) protested about the loss of biodiversity and entrepreneurial piracy of farming knowledge when multinational firms treated local farmers in India as objects from which information could be mined to benefit shareholders based overseas. Chileans have opposed the construction of dams in the fragile ecosystems in the wilderness of the Patagonia. Local people in Turkey have protested against the loss of cultural heritage in Hasankeyf, despite government assurances that the proposed construction of the Ilisu Dam to produce hydroelectricity will be environmentally sensitive, although it will raise water levels by 200 feet and destroy ancient cultural artefacts (Bolz, 2009).

These protest actions highlight the significance of probing beneath the surface of decisions that may appear 'green' to assess the full impact of each and every proposal to change the environment for purposes articulated by people with capital funds to invest and governments responding to social needs that do not reflect local concerns or answer questions about who benefits from such schemes. Building hydroelectric dams is a case in point. Constructing these does have a social cost which is carried by the communities and biospheres sited on land given over to their construction, but this is neglected by developers (Roy, 1999). Although not an exhaustive list, this includes the dispersal of people from ancestral homes with inadequate compensation (if any) so that these can be flooded to build reservoirs, and all this entails in terms of: destroying a sense of belonging and ancestral heritage; the removal of the original community as a result of the mass relocations of people; the disappearance of traditional livelihoods; the destruction of specific ecosystems; and loss of local knowledge about an area, its peoples, flora and fauna.

Environmental impact assessments currently in use draw on economic cost–benefit analysis which attaches a quantitative figure to the basic costs of providing materials and labour to an initiative, e.g., building a dam (ECA, 2005). These are not holistic in their calculations. Even in situations where the costs of industrial development to people are acknowledged, the formula utilized provides a simplistic mechanism that assumes that technological innovations, the replacement of natural resources by manufactured ones and a more efficient use of resources will solve any problems that might be encountered, including any shortages in natural resources that might ensue. It pays no attention to emotional attachments to place, current endeavours in eradicating poverty among poor and marginalized populations, safeguarding resources for future generations, the long-term damage caused to the physical environment by the act of development itself or unintended consequences of such initiatives, e.g., the advent of drought downstream of rivers that have been dammed for power generation. Some sustainable development models have successfully integrated environmental and economic considerations, without incorporating the social justice issues of distributional and procedural equity or quality of life considerations essential to human and environmental well-being (Agyeman and Evans, 2004); and the holistic protection of the environment. A few examples of social entrepreneurialism embedded in notions of the social economy have sought to address the matter of producing goods and services required by people while ensuring that economic development supports marginalized groups.

Nonetheless, many of these have tended to ignore environmental concerns linked to the physical environment (Amin et al., 2002).

In the building of the Three Gorges Dam in China on the Yangtze river in Hubei Province, at the cost of 254 billion yuan (£24 billion), 1.5 million people were forcibly evacuated for the greater good. Despite this compulsory or enforced migration, current outcomes illustrate the inadequacy of existing approaches to traditional cost–benefit analysis and environmental impact statements. The Three Gorges Dam is the largest hydroelectric project in the world. It was the experts, not the local people, that determined who would bear the cost of the resultant disruption when 13 cities, 140 towns, 1,350 villages and 1,300 archaeological sites were submerged to build the Earth's largest reservoir. Pollution, soil erosion and silting have plagued its success. Now, it is deemed to be causing soil erosion, drought, food shortages and social upheaval as farmers protest against what has been happening since it opened (Watts, 2011a).

Meanwhile, a drought in the middle section of the Yangtze river has resulted in 1,392 reservoirs in Hubei Province holding only 'dead water' and destroyed access to clean drinking water for 300,000 people. And dropping water levels near Wuhan have meant that boat traffic has been banned from the river for lack of water. It is not surprising that Chinese people have protested about this situation before the dam was built and since; a recent protest occurred in early 2011 (Watts, 2011a). The Chinese government has finally acknowledged these difficulties, but it continues to emphasize the gains brought about by its construction, especially the availability of electric power in poor areas. It also promised more measures to correct these problems, e.g., providing an early warning system so that people could take precautionary measures, including evacuation in good time, and improving benefits for the additional people it thinks will have to be relocated to reinforce the riverbanks. It has also sought to restore the ecosystem, although this has been difficult because, once altered, it cannot be the same again. Additionally, the project is in a seismically active region, as is much of China. A question has to be asked about whether smaller and environmentally driven projects in which prestige is not at stake may be more environmentally friendly and sustainable in the long run. These issues should have fully informed the balance sheet that was totted up when the initial assessment of the impact of the Three Gorges Dam on the environment was made.

Social workers can mobilize people and assist them in developing more holistic, environmentally friendly, co-produced solutions to projects such as these, and endeavour to raise a different kind of consciousness among government officials, developers and multinational firms involved in their construction. In Turkey, community residents have proposed locally based, alternative sources of power as the way forward (Bolz, 2009). Developing counter-proposals in China is more difficult. The sheer size of the population and areas affected are significant barriers, and the government's imperative to promote the country's economic development over-rides other concerns, despite its commitment to alternative energy technologies. An interesting aspect of social and community development in China is that community workers have to be extremely aware of local cultures, languages and traditions among a varied population, and work carefully with local officials to promote what they call 'harmonious' communities, and operate within government plans for an area (personal communication). The lessons of the Three Gorges Dam must be learnt and applied in the proposed construction of significant hydro-generating capacity in other locations, e.g., the Nu river.

Aung San Suu Kyi, the elected pro-democracy leader of Burma/ Myanmar deposed by a military coup and held under house arrest following her election, has protested against the proposal that China Power Investment builds and manages eight dams along the Irrawaddy river in a number of projects that include the Myitsone Project, which will be built in a seismically active area, at a cost of US$3.6 billion. The one to be built at the confluence of the N'Mai and Mali rivers in Kachin state will flood an important rainforest and displace 10,000 people so that the Burmese government can sell electricity to other countries on its borders, particularly China and India, with sales estimated to reach US$500 million annually. Aung San Suu Kyi labelled the Myitsone plan 'dangerous and divisive' and called for a re-assessment of the project and dialogue among interested parties. Local people have complained that it will damage their environment, cultural traditions and social fabric. The Kachin Independence Organization has supported their concerns. A leaked environmental assessment suggested that the project should be scrapped because enormous environmental and social damage would occur in the event of an earthquake (Watts, 2011b). Already, a number of people have died in clashes over its construction. Such complex issues emphasize the importance of an unhurried and inclusive approach to development that involves all stakeholders, especially the local people who will be most affected and who have a specific stake in caring for the environment as a living entity for current and future generations, not just company and government representatives. All stakeholders ought to have an equal share in the benefits and all aspects of the work and investment sums involved must be transparent. Social workers can assist the Chinese government in taking these complexities into account as it has enacted a moratorium on the Myitsone development, although local people question whether the cessation of construction has actually materialized (Forbes, 2011).

In Africa, the damming of the Zambesi river to create the Kariba Dam as long ago as 1961 has disrupted the flow of water over Victoria Falls, and impacted badly upon the 57,000 Tongo people whose communities were relocated and families divided between what is now Zambia and Zimbabwe, making visits between family members difficult as border crossings are involved; endangering local ecosystems; and exacerbating environmental crises, so that the region currently suffers from drought. Enforced migration or internal resettlement in this case, as in the Three Gorges project, does not necessarily solve problems. Famines occurred in resettled areas because the agricultural lands provided to the Tonga people were so poor (Scudder, 2005). Social workers intervene to support resettled people and those who are starving in obtaining access to food, water and medicines, and counsel people who have been traumatized by the experience. Such interventions also have to address issues of compensation for lost property, sense of belonging and heritage.

In the UK, the flooding of Welsh valleys to meet the water needs of English populations living in large urban centres became a sovereignty issue because those living in Wales lost amenities that future generations might need for themselves. For example, the building of Lyn Celyn, or the Tryweryn Reservoir, to provide Liverpool with water necessitated the flooding of Capel Celyn and other parts of the Afon Tryweryn valley in the mid-1960s. Imposing its construction on a reluctant Welsh population and passing an Act of Parliament that allowed this development to proceed despite opposition from Welsh MPs and their constituents, contributed to the rise of the Welsh National Party, Plaid Cymru. An apology for the specific appropriation of these resources, bypassing Welsh local authority jurisdiction and neglecting of Welsh sentiments on the matter was not forthcoming from Liverpool City Council until 19 October 2005 (*BBC News*, 9 March 2006). Social workers could assist people in advocating for their right to have their voices heard and concerns addressed.

These examples illustrate various complexities and tensions that suggest a more holistic understanding of issues involved in degrading the environment for developmental purposes is necessary. These should encompass the socio-cultural, spiritual, political, economic, biological, material and physical spheres. And they should highlight the inadequacy of economically determined cost–benefit analysis and reverse the privileging of the economic sphere over the social, as is presumed in traditional economic analyses of such matters (FNEATWG, undated).

Questions about the economistic and quantitative orientation of the assessments of the impacts of ecological change on people and their environments have to include the organization of work and payment of wages or salaries to those providing the labour that underpins economic growth and the profits that companies extract. These concerns indicate the importance of demands for structural change in how society: manages social relationships, including workplace relations; distributes power, wealth and resources within it; and organizes wealth creation, commodity production and consumption, and environmental protection. Current social organization privileges those with money and leaves large numbers of the world's population in poverty, marginalization and social exclusion. These are issues that green social workers are concerned about at local, national and international levels because they carry implications for human well-being and the environments in which people are situated.

Social workers can be involved in developing holistic assessments of situations that would include environmental concerns, and assisting people in compiling a case for the reform of capitalist social relations. A call for the reform of capitalism is rarely found in mainstream discourses about development and/or sustainability, and politicians prefer to focus on reforming the welfare system, if there is one, rather than this fundamental issue which would require creative innovative thinking on their part instead of repeating the well-rehearsed refrain of modernizing welfare and lowering people's expectations about being cared by others through the solidarity embodied in the welfare state. The entitlement society, not its economic underpinnings, is defined as the problem in this view of the world. It also continues to ignore the consequences of retrenchment and neoliberal welfare policies on people's lives (Allan and Scruggs, 2004).

A reform of capitalist social relations would include eradicating social exclusion and the unequal distributions of wealth, power and resources, and respond to people's demands for alternative forms of production, consumption and reproduction, based on different worldviews that are predicated on social justice, human rights and citizenship-based entitlement to goods and services. The current failure to reform the gross inequalities perpetuated by capitalism in Greece provides ample evidence of this need. An alternative that addresses people's demands to have jobs that provide decent rates of pay and care for the environment challenges the profit motive and market-based economic priorities by widening the circle of factors to be taken into account, so that they encompass those based on service, equality, solidarity and reciprocity. Sustainable development and initiatives like social enterprises are growing in popularity because they aim to respond to these concerns. However, caution is advised. These ventures have not usually replaced the profit-motive or market-driven discipline as the major forces behind the economy. Rather, these social entrepreneurial schemes exist within the interstices of the market-place, and their positionality as alternative economic structures and action as transformers of existing social relations may be blocked as a result.

Migration as a Response to Social, Economic and Environmental Crises

Migration has been a phenomenon permeating human history from time immemorial. The actual numbers of people involved are difficult to calculate. In 2010, there were millions of people moving around the world for various reasons. Their migration has been motivated by a range of 'push' and 'pull' factors (IOM, 2010). Among the 'push' factors are: the lack of employment opportunities; natural disasters that have destroyed food crops, homes and income generation activities; and armed conflict over land or other resources for nationalistic reasons that exclude some people's claims to legitimate ownership of land. The contested nature of land claims, especially long-term ones, leading to armed conflict is exemplified in the Palestinian territories where Palestinian people holding deeds to their lands find them occupied by Israeli settlers who are guarded by armed soldiers preventing them moving past check-points into what they deem is Palestinian property. Israeli state policy has been dubbed 'aggressive urbanization' as olive groves, other plantations and trees are uprooted to make way for settler residences (CCDPRJ, 2009). These issues are complex, and not within the remit of social workers to solve. Practitioners can become engaged in endeavours aimed at facilitating dialogue across fundamental differences and working with those involved to find non-violent ways of resolving disputes and developing relationships and networks that will build resilience and enable them to survive tough times. Social workers have assisted in forming groups, especially women's groups, to promote dialogue between Israeli and Palestinian women in order to achieve peaceful reconciliation. In other examples, practitioners residing overseas have linked up with groups in conflict to encourage dialogue between the disputants and provide services wherever possible. In one instance, Norwegian social workers have facilitated exchanges with Palestinian social workers through the Palestinian Union of Social Workers and Psychology (PUSWP) (Sturge, 2010). The British Association of Social Workers has recently joined this endeavour. They aim to work together to support Palestinian social workers in whatever ways they request, e.g., exploring training opportunities, developing culturally relevant curricula and facilitating access to the wider social work community. These illustrate possibilities that can be developed further and expanded to include others. However, progress in such initiatives has been slow.

Migration caused by human-induced climate change is a relatively new phenomenon. The numbers involved are disputed, but are predicted to vary between 200 million and 1 billion from 2010 to 2050 (Myer,

2005; Christian Aid, 2007). Such migration will not be evenly spread. Despite the impact of climate change on all residents in a community, women working in agriculture are less likely to migrate for climate change reasons than men (Massey et al., 2007). The differentiated impact of climate change raises interesting issues about social and environmental justice. It is not usually people whose industrialized patterns of development and consumerist lifestyles have precipitated climate change who are most likely to be affected. Thus, it is crucial that equity issues, poverty eradication and the interests of marginalized groups are centre-stage in proposed solutions. When considering these issues in this chapter, I explore the roles that social workers have played and can play in these circumstances.

Migrating in search of food

Nomadic peoples have migrated in search of food for millennia, following the seasons in foraging for food and water for themselves and their animals. These ancient patterns and rhythms have become disrupted through prolonged drought, the damming of traditional waterways and desertification. Whilst these calamities may be natural to some extent, the devastation they cause can also be exacerbated by human interventions including forms of armed conflict. What may begin as a natural disaster, e.g., drought, can become even more hazardous through people's actions as they pursue other political, socio-cultural and economic agendas. Armed groups are often integrally involved and can block aid from reaching intended recipients, the victim-survivors. Or they utilize food aid given by others for their own political purposes by adding these resources to local ones already in place (Abdisaid, 2008; Escobedo, 2009). In 2011, the African Union Mission in Somalia (AMISOM) claimed that Al-Shabaab (Ash-Shabaab, Hizbul Shabaab) is blocking aid to starving people in Somalia and preventing their exodus to relief camps on the grounds that any help they would receive would not meet the requirements necessary for leading an Islamic life. According to AMISOM, an Al-Shabaab member drove a truck bomb into its base in Mogadishu killing twenty-one peacekeepers and himself. These activities inject new hazards into equations of humanitarian aid that are of a political nature, and they involve people making decisions about or for other people. I term the complex relationships involved in these manoeuvres the *politics of disaster*. They require political solutions at local, national, regional and international levels, rather than involving social workers in local-level negotiations around the neutrality of their humanitarian intentions. Social workers seek to bring about reconciliation by trying to promote dialogue between different fractions of the population, even in war zones. Some lose their lives in the process. The players actively involved in conflict situations may not recognize the politics of disasters because they normalize their decisions and see these as reflecting the usual routines of everyday life in conditions of armed struggle, including liberation struggles. Such events are becoming the norm in the Horn of Africa, for example.

Some environmental issues are exacerbated further by colonial legacies. In one instance, tribal peoples living along the borders of Kenya, Somalia and Ethiopia were prevented from moving across the borders by conventions created by the European colonization of Africa. This is illustrated by the Dadaab camp in Kenya, which was created initially for Somalian refugees following the collapse of the Said Barre dictatorship. Although social workers as relief workers have established refugee camps for any displaced population, government regulations stipulated that those of Kenyan nationality could not use these. The Kenyan government's demand that its nationals desist from entering Dadaab camp on pain of losing their citizenship rights did not stop Kenyan nationals affected by drought from seeking support there. The government's response raises concerns about the arbitrariness of the frontiers that colonial powers imposed on local nomadic people. And it poses the question of why people should lose their rights as citizens when asking for help from their national governments simply because they cross over into lands that they had traditionally used as herds-people. The numbers of people that descended upon this camp from the three adjacent countries suffering from drought – Somalia, Ethiopia and Kenya – exceeded capacity. Double the 90,000 people that the camp was able to cater for arrived in 2008–9. Pressures from these numbers intensified environmental degradation. Their populations rose dramatically again as a result of the enduring drought that precipitated a further crisis in 2011, when 400,000 people were living in Dadaab camp by July and extension camps had to be developed. Moreover, the food provided to refugee camps through the World Food Programme has been insufficient. Similar problems arose in other camps, e.g., Ayub in Ethiopia. Working in such conditions is also very stressful for the practitioners involved.

Recurrent famine in the Horn of Africa

Population growth and food security are inextricably linked. Population numbers are expected to rise, but these will not be evenly distributed across the world. Estimates indicate that between 2005 and 2050, the nine countries of India, Pakistan, Nigeria, the Democratic Republic of the Congo, Bangladesh, Uganda, the United States, Ethiopia and China will be responsible for half of the world's projected increase in population. China's contribution will become greater if it ends its one-child policy. Population increases through in-country births in Western countries are expected to remain low, while out-of-area immigration is likely to increase.

Population numbers will also rise as life expectancy does. Life expectancy at birth for the whole world has risen from 46 years in 1950 to 65 years by 2005. Projected increases suggest that this average will exceed 75 years by 2050. Increased longevity will be unevenly distributed within countries and between them. In the industrialized West, life expectancy is likely to rise from 75 years today to 82 years by 2050. Within countries, regional variations and fracturing along other social divisions will occur. For example, life expectancy for men in London, England, is 84.4 years, compared to 71.1 years for Scotland's Glaswegian men. The comparative figures for women are 89 years compared to 77.5 years. Life expectancy in poorer industrializing countries is currently under 50 years, but this is projected to rise to 66 years by 2050 (UNESA, 2009).

Current projections suggest that, by 2050, Africa will be home to 2.9 billion people, Asia to 5.2 billion, Europe to 674 million, Latin America and the Caribbean to 765 million, and North America to 448 million. At that point, the projected number of children per woman globally is expected to fall to 2.05, with the Middle East and North Africa having 2.09 and sub-Saharan Africa 2.61. This reflects a considerable drop for all regions except Europe, which currently is not replacing its population without immigration (UNESA, 2009).

Population is also growing rapidly in countries experiencing environmental crises. For example, Ethiopia's population grew from 35 million in 1984 to 85 million by 2009, without a commensurate increase in agricultural production. Most of its population lives in rural areas where farming continues to be undertaken according to pre-industrial modes of production. Thus, the country is unable to produce sufficient food for current population numbers. The famine of 1984–5 in Africa was severe and affected primarily Ethiopia. It encompassed large numbers of poor people, and was exacerbated by armed conflict involving the Derg dictatorship and those fighting for the liberation of Eritrea and Tigray. This famine generated a crisis of historic proportions and claimed over 1 million lives. The international response was so inadequate that Bob Geldof, a pop star, and others created Band Aid and put together a Live Aid concert to raise the profile of the disaster and give attention to its victim-survivors, and raised £150 million for relief purposes. Although not social workers, they performed social work tasks like responding to people urgently needing help. It illustrates the ease with which non-social workers can appropriate social work's remit. At the end of the day, relief workers are the ones to transform these contributions into goods and services that reach starving people on the ground.

This assistance was crucial in preventing many deaths. Yet, it was an insufficient response. Aid in the form of food handouts has been criticized, including by children, who are now adults, fed to survive Ethiopia's 1984–5 famine. Whilst those suffering need immediate relief, especially water, food and medicines, Ethiopian survivors of this famine, like Birhan Woldu whose photo as a starving three-year-old was used by the media to secure donations for the cause, argue that such responses encourage dependency, and that addressing the structural issues in agriculture and enabling Ethiopians to grow their own food would provide the more effective longer-term solutions needed. This view has been encapsulated by the motto of 'don't give me fish, teach me how to fish'. On a global level, 7 per cent of aid is spent on food aid.

In Ethiopia, 91 per cent of aid in 2009 was used to supply food. This fed 6 million people, half of them under the age of eighteen. A succession of droughts, including several very serious ones occurring between 1984 and 2011, had brought millions to starvation a number of times. Those as recent as 2003, 2008 and 2011 were particularly devastating. Meanwhile, the Ethiopian government has sought to mitigate the impact of a potential famine by: funding the Productive Safety Net Programme; increasing benefits for those suffering from drought; developing a Famine Early Warning System that monitors rainfall, livestock prices, household spending and malnutrition; stockpiling food; increasing spending on agriculture which comprised 17 per cent of government expenditures in 2003; forcibly resettling 600,000 villagers from drought-stricken northern regions to the south of the country; and improving the planning system. These measures have still failed to prevent mass starvation. Additionally, many of those resettled returned to their former homes, even though these were in areas that experienced armed conflict, thereby highlighting the importance of ancestral space.

The inadequacy of humanitarian responses indicates the importance of responding holistically to such disasters. It is not only agricultural policy that needs to be addressed, but also urbanization, cultural heritages, economic development, political stability in contested regions, and population growth. In August 2011, the Horn of Africa was officially designated as facing another famine. UNICEF suggests that, this time, the numbers involved in this region encompassing Somalia, Kenya and Ethiopia, had risen from 8 million in 1984 to 23 million. The UN defines a famine as a situation in which 20 per cent of households face extreme food shortages, acute malnutrition exists in 30 per cent of the population and 2 deaths occur per 10,000 people per day. Much of the current crisis is centred on Somalia, where peoples endured another devastating famine in 1991–2. The current food crisis threatens to exceed the numbers reached in 1984 as it spreads. Deaths among children less than five years of age had reached 13 per 10,000 per day by mid-2011. Getting relief aid into the country is compromised by Al-Shabaab's control of the area and blockage of foreign aid destined for people living within it from reaching them. Yet, the price of bread has doubled in the south of the country where refugees are going in substantial numbers (Martell, 2011). By the summer of 2011, the UN's Food Security and Nutrition Analysis Unit for Somalia estimated that 100,000 refugees seeking water, food, medicine and shelter had reached the country's capital, Mogadishu. These numbers are swelling by over 1,000 daily (Martell, 2011). Another 400,000 Somalis have fled into the Afgoye corridor in search of food aid, and their numbers have made this the site of the largest camp for internally displaced people in the world.

Responses to the UN's international call for funds to assist those caught up in this disaster have been slow. Only half of the US$2 billion required had been committed by August 2011. The USA, the largest single donor, had offered US$450 million. The Canadian government pledged to match private donations to registered charities over a 10-week period, beginning 6 July 2011. Some authors have termed the lack of response to calls for continued donations 'compassion fatigue'. For me, this highlights the inadequacy of a relief system that asks for individual, organizational and governmental donations on an ad hoc, voluntary basis as the way of supplying life-sustaining resources that enable people to survive in the short term. I would want to argue for a mandatory set-aside of a relief tax levied on the earned income of all citizens of the world to help each other in time of need. This money would be collected by national governments through normal taxation mechanisms, but ring-fenced to support immediate relief and long-term reconstruction following disasters. The relief tax would be placed in a Disaster Solidarity Fund (DSF) and divided equally between national-level emergencies and the UN for international relief. Ring-fencing should help governments maintain transparent records that should minimize its use for other purposes. Social workers embedded in local communities could assist local residents in monitoring such funds and compiling records of whether the aid purchased is distributed, whether it reaches the appropriate people, and how it is collected and used on the ground.

Contemporary disasters show that no one country is immune from their impact and in arguing for a solidarity-based, interdependent and egalitarian world, it is important to find reliable, non-stigmatized sources of aid for short-term relief, and longer-term development money to eliminate poverty and develop the necessary infrastructures and organizational mechanisms needed to ensure that no one person is excluded from enjoying a decent standard of living. Developing resilient resources for all is a matter of political will, and of enabling everyone to access these as needed. It also requires a fundamental rethink of emergency responses as currently practised. Social workers have an important role to play in highlighting weaknesses in current relief systems and advocating for more sustainable responses that are embedded in and owned by the communities concerned. Local systems should be resilient enough to tap into resources provided nationally and internationally. Overhauling the current relief system requires solidarity that: recognizes interdependencies between people and the natural environment; encompasses social justice, environmental justice and human rights; and supports an obligation to care for others and the environment and the right to be cared by others and the environment, as well as a responsibility to care for both. Research is also required to develop robust systems, and this should involve a wide range of experts who could work effectively with local people to co-produce the solutions appropriate for the locale. The ring-fenced taxation scheme or DSF described above could be one way of getting the finances for communities to move forward in a financially sustainable way.

Environmental Degradation and Food Production

Population growth has considerable implications for food production and consumption and environmental degradation. Until the Green Revolution increased food yields so that food production could keep pace with population growth, the balance between them was an uneasy one. This advantage was undone to some extent by the environmental damage caused by nitrate- and phosphate-based fertilizers. De Moor and Calamai (1997) argue that this gain has been lost because technological innovations carry deleterious environmental costs that eventually caused soil erosion and a decline in food production over time. Consequently, current population growth globally could see the return of conditions in which population size seriously outstrips food supply, and inadequacy in the quality and amount of food available carries with it malnutrition, hunger and famine across more parts of the globe.

Environmental degradation carries implications for all aspects of life. The pollution of land, water and soils, desertification and soil erosion can undermine food production on two levels: pushing out small producers; and increasing reliance on externally produced foods. The spread of industrial mass-produced food through agribusiness has deleterious consequences as people often experience ill-health, including respiratory ailments, from living in a polluted environment and/or obesity caused by eating fast food with high levels of calories, carbohydrates and transfats. These outcomes raise questions of equity because it is mainly low-income populations that consume health-poor foods by ingesting products that are mass-produced, cheap, but nutritionally of poor quality compared to health-rich foods, as exemplified by the more expensive organic, locally grown fruit and vegetables that are rich in vitamins, minerals and other nutrients essential to good health (Schlosser, 2001).

The mechanization of agriculture and need to feed large numbers of people living in urban areas encouraged the rapid rise of industrial, mass-produced food. This successfully reduced uncertainty of supply, but could produce irreversible damage to the environment through the chemical ingredients used to grow food rapidly and in sufficient quantities. Although environmental degradation was often an unintended consequence, chemicals utilized in fertilizers seeped into groundwater; and those used in herbicides could compromise soil texture and the ecosystem (Carson, 1962). Carson's book *The Silent Spring* is credited with the banning of the pesticide DDT (dichlorodiphenyltrichloroethane) in the USA in 1972 by exposing its deleterious effects on human health and the physical environment. In the industrialized production of food scenario, manufactured capital used in producing food replaces natural capital in food production and consumption, thus promoting the drive for technical innovations that transcend the limits to growth. The Green Revolution that enabled humanity to increase food production to unprecedented levels so that now, at least in theory, enough food is produced to feed all the Earth's inhabitants is the classic example of this. However, food is not equitably distributed, and so billions of the Earth's people continue to suffer from hunger. Such developments challenge neither the dominant hegemonic mode of capitalist production, nor its structural inequalities based on 'race', gender or other social divisions that intensify poverty and marginalization and contribute to some people lacking food while others have a surfeit of it. Indeed, poverty is integral to capitalist systems of production (Marx, 1978). At minimum, capitalism's excesses have to be curbed through the principles of equality and solidarity, and, increasingly, valuing the environment. Some initiatives promoted by community groups in the social economy movement grew out of a desire to respond to peoples' needs while valuing both people and the environment.

Connelly et al. (2011) argue that environmental considerations and social justice issues can be addressed through sustainable initiatives rooted in the social economy. For this to occur in food production, just and sustainable local food systems have to become part of local community development and grow to establish an infrastructure that can transform existing food systems into more environmentally friendly and socially just ones. This requires the mobilization of 'citizens and governments through democratic processes [that] . . . coordinate, balance and catalyse the values, visions and activities of various community actors to create change' (Roseland, 2005 quoted in Connelly et al., 2011: 310). This venture is difficult to achieve because social enterprises exist within a non-supportive macro-system of profit-driven production. Some argue that social enterprises are incapable of generating their own capital but rely on subsidies from the mainstream economy and the politics of redistribution (BoE, 2003).

The politics of redistribution have lost significance in the political repertoire of the nation-state in favour of 'business as usual' models that prioritize economic needs over social needs. This makes poor people bear the brunt of reduced public funding for welfare services free to all at the point of need, and includes cuts to subsidized foods such as bread, milk and grains. People-centredness, the integration of the personal and structural, not the subordination of the personal and 'social' to the for-profit neoliberal economy, is required to secure structural and collective action that reasserts the politics of redistribution whereby social and economic resources and political power are equitably distributed and social relations altered to make this possible. Reconfiguring professional relationships to empower service users by enabling them to design and deliver services and validate their knowledge of everyday life practices is crucial to the processes of transformative change (Dominelli, 2010b).

Food cooperatives and other forms of collective action aim to expand alternatives and transform the production and consumption of food, including the development of short value chains between consumers and producers (Winter, 2003). These alternatives are embedded in locality, or local sites that produce natural, healthy food through systems predicated on reciprocity, trust, transparency and accountability. These have to find avenues through which food security is ensured and include those unable to pay the full environmental, social and economic costs of its production. Such endeavours aim to engage in the politics of redistribution, but success in reaching this goal may be difficult because alternative systems currently have to operate within the interstices of a hegemonic neoliberal capitalist framework until it is transformed.

The development of alternative food systems is affected by changes to land use, especially in rural areas. This includes the urbanization of the rural environment as developers seek 'green field sites' in preference to 'brown' ones, because the former are cheaper to develop and make a profit from. Economic costs do not include the costs borne by the individual consumer who has to add transportation, cooking, labour and other charges to the price paid to the cashier. This individualized cost does not include the environmental costs which are carried by the physical environment in all its aspects, nor does it cover those costs accrued by plant and animal species. This may even mean their demise. Planners and politicians often fail to take this wider view of costs and agree to redevelopment plans that are socially predatory and environmentally unsustainable.

Other issues for alternative food systems are long-term sustainability, funding associated with it and expansion beyond the immediate locality to keep costs low and counter problems engendered by a generally hostile financial climate. Some hostility towards the mainstream ecological movement arises from its objective of de-globalizing, de-industrializing and de-scaling food production and consumption infrastructures that typify agribusiness (Goodman and DePuis, 2002).

Case study

The New City Market Local Food Hub (NCM) was developed by local food organizations to provide food among poor people living in Vancouver, Canada. The NCM was created by multiple stakeholders in the Local Food First (LFF) initiative that began in 2005 to create a locally sustainable food system and debate issues linked to how food was produced, costed and distributed. LFF was involved in action and analyses that aimed to: rebuild the local food system; redevelop local food value chains that enabled farmers to have direct access to an expanding local food market, pay fair wages and create good working conditions for workers; regenerate the local food infrastructure; and ensure food security. The formation of the NCM signified the fruition of these goals. Much work remains to be done. This includes finding potential sites for expansion; exploring different business and governance models; and securing capital for its endeavours in food distribution, warehousing, cold storage, and for other requirements (Connelly et al., 2011).

In Vancouver, some developers provided space for the formation of a Local Food Hub in their plans. This may seem an altruistic gesture, but it carries the danger of distorting the values, objectives and power dynamics that drive alternative food systems. These include the self-determined location of such sites; revenue generation; capacity to act as pathways to community transformation; and the formation of new social capital (Hanson, 2009). All this can occur if appropriate conditions are set on how these systems operate, e.g., not being concerned mainly with making a profit.

The NCM advances Vancouver's policy on developing sustainable local food production. However, such alternatives are required to comply with conditions ensuring business viability and to provide business plans that demonstrate that the initiative is financially viable. These demands, based on traditional models of business planning in neoliberal societies, can undermine NCM's commitment to social justice. This may happen because traditional models do not approve of businesses taking risks by providing food below costs to some groups to engage them as actors in market situations, while mitigating the risks posed by spreading the actual costs to those able to pay more. These tensions highlight the difficulty in transforming local food systems away from the hegemonic industrial, mass-produced profit-driven ones and asserting the importance of alternative systems to local people unable to purchase healthy food in the market-place. Nor does the industrial model ensure that both producers and consumers are involved in making decisions about environmentally sustainable food production and consumption.

Alternative food production, according to Dixon (2011) does not have to be anti-capitalist. But it does ask for accountability from food producers, the involvement of local people in food production and consumption, and the distribution of food to everyone that needs it. Social workers as community development workers play important roles in linking various stakeholders, including farmers and community residents; coordinating activities at different levels; promoting collective action in meeting the food needs of marginalized groups; and fostering the development of trust among different players involved in the diverse stages of bringing alternative food systems into being through the formation of food cooperatives and related initiatives, such as the NCM. Bringing these into existence and securing their future remain challenging tasks for social workers, as community development workers, to realize.

Holistic disaster interventions

There are a number of different natural disasters that afflict humanity and the planet. Social workers can support people in the different stages of disasters by helping them to: understand their nature; mitigate risk by knowing about their potential to cause damage; take steps to prevent disaster; minimize the damage caused; and address its consequences. Below, I consider the impact of crucial 'natural' disasters on communities. A holistic approach to problem-solving is handy in tackling need in such situations. One of these natural disasters was the heat wave of 2003 in Europe that caused the deaths of thousands, especially among older people with compromised immune systems. Estimates on the numbers of people losing their lives ranged from 35,000 to 50,000, including 15,000 in France, 7,000 in Germany, 4,000 in Italy, 4,000 in Spain and 2,000 in Britain (Larsen, 2003). The death toll marked this heat wave as the worst recent 'natural' disaster in Europe. A key social care issue to be addressed by social workers is that of ensuring that older people and children keep cool and drink plenty of cold liquids during heat waves. This is to prevent possible dehydration of their bodies in the heat because this can cause their body temperatures to rise to dangerous levels and precipitate heart attacks or death. Following this heat wave, a national warning system that involved social workers was created in Spain to prevent deaths in future events. Similar plans have been developed in other parts of Europe (WHO, 2004; Kovats and Ebi, 2006).

Other heat wave dangers are havoc caused to crops by drought that endangers food supplies. Reducing such risks necessitates the protection of food supplies and ensuring that vulnerable people have access to these. Social workers realize these roles by distributing food, drinks and medicine in safe centres or in people's homes. Providing home-based assistance for all those needing it requires social workers to promote preventative srategies at the level of the community before disaster strikes, linking up with emergency planning teams that determine which responses to disasters would be most effective and developing a network of community volunteers skilled in disaster intervention techniques that they can call upon should disasters occur. They also have to link up with farmers, agriculturalists and engineers and input the social dimensions of heat waves and their impact upon people into discussions about food security.

A network of trained, well-informed community-based volunteers would also be useful if forest or wildfires coincided with heat waves. Portugal, for example, lost 40 per cent of its forests during the hot, dry summer of 2003 and could have used such volunteers to complement the work of overstretched fire-fighters and other emergency professionals. Wildfires occur regularly in Canada, the United States and Australia. These countries could also use a network of already-trained volunteers living within wildfire-prone communities to mitigate the risks of such events occurring, having them ready for action should they be needed. Their training would have to occur before disaster strikes and could provide the basis for a national, community-based service (NCBS) aimed at helping people and society as part of their normal life activities. This would be different from the type of community service that Prime Minister David Cameron is proposing in the UK for offenders. His scheme is aimed at controlling people because community residents might engage in behaviour that poses threats to the social order. An NCBS helper would always be on standby and require constant training and upgrading of skills to cater for locality-specific disasters. Such volunteers would have the advantage of knowing local languages and cultural traditions.

Other dangers are less easy for social workers to address, although they can alert communities to potential dangers before there is an emergency. For example, as rivers dried up during the 2003 heat wave, nuclear reactors were endangered as there was insufficient water to keep them cool. Cold water is required to ensure that water discharges from individual reactors remain below 25°C. Temperatures inside the plant must not to exceed 50°C. To maintain reactors below this temperature, 200 million litres of cold water have to circulate through each one each day. During the 2003 heat wave, the nuclear plant at Fesenheim in Alsace reached 48°C. Engineers sprayed the concrete casing outside it with a fine water spray derived from groundwater to keep it cool 24 hours per day. This action lowered the temperature by 1°C. This outcome was worrying, given the potential threat had air temperatures continued to rise. Thankfully, the heat wave ended before it produced a major energy emergency as occurred in Fukushi Daiichi in Japan in 2011 after the tsunami destroyed key components in its cooling system.

Another concern in Europe revolves around melting glaciers causing higher levels of water in alpine rivers that threaten to flood villages and towns in Central Europe. Social workers become involved in helping people cope with the aftermath of flooding once they are moved to places of safety. Some also assist in evacuation efforts to get people affected by disasters to safe havens. But there is more they can do, including helping to create an NCBS. Lessons learnt from previous natural disasters help policy-makers and emergency planners improve interventions in subsequent events. Spain, for example, established a special register whereby those most at risk in times of natural disaster could sign up for special services. Social workers were envisaged as a key professional group ensuring that these services were created and delivered when required.

Russia faced the reverse problem during a severe cold spell in the winter of 2006, when temperatures dropped to −30°C. Here, people froze in their homes when they lost electricity and central heating for prolonged periods. A register of those at risk in extreme weather events, like that devised in Spain could have identified people who would have needed help and ensured that services reached them in time, especially if there had been an NCBS to draw upon. The plight of people at risk and their numbers raises concerns about the substantial amount of work for social workers to handle during extreme weather events, and their capacity to take it on. It highlights the need for additional labour power that could be made available to them through an NCBS system. Additionally, the profession is poorly equipped to handle these calamities on its own, and so it is important that it develops links with physical scientists and other emergency professionals involved in disaster interventions to acquire the relevant scientific knowledge and ensure that the community co-produces the solutions required.

Tornadoes provide another 'natural' disaster with victim-survivors whom social workers help. Tornadoes have traditionally been measured through the Fujita Scale, which roughly calculates wind speed in a measurement scale ranging from F0 to F5. In this, wind speed rises as the higher numbers in the scale are reached. Wind speeds of between 40 and 73 mph (64–117 km/h) at F0 move up to reach 260–318 mph (418–512 km/h) at F5. Fujita Scale measurements were deemed inadequate because they did not assess damage to structures along with wind speed. In 2007, the Fujita Scale was replaced by the Enhanced Fujita Scale, which assesses damage to structures along with wind speeds. Practices have also changed with lessons derived from past disasters. For example, before the mid-1970s, the weather service did not issue warnings about oncoming tornadoes or other extreme weather events as a first line of defence. Now it does. This is to give people time to take preventative measures like evacuating an area and making homes, possessions and families as secure and safe as they can.

Desertification is another serious environmental issue. Some places are experiencing more problems in this regard than others, e.g., Africa, Mongolia. In China, the desert is creeping up on Beijing and is now a mere 43 miles (70 km) away. The situation is aggravated by sandstorms that remove millions of tonnes of topsoil and shift it from one area to another, often thousands of miles away. On 17 April 2006, one of the worst sandstorms on record reached Beijing from Inner Mongolia. In another instance, in April 2001, the dust cloud from Inner Mongolia travelled across to North America, taking six days to traverse the Pacific Ocean. The loss of millions of tonnes of topsoil creates dustbowls and respiratory problems for people caught in its path. This stresses hospital and social care provisions because demands for these services can increase dramatically during such events. And they may be ill equipped to respond. These difficulties are likely to increase, given that China has one of the highest erosion rates and degrees of environmental degradation in the world and current government-led endeavours have failed to resolve these problems.

Deserts now claim 20 per cent of China's landmass. Initiatives such as the planting of the 'Green Wall' of 35 billion trees planted to hold back desertification have not had the anticipated successful outcomes. Some scientists argue that this strategy was ill-conceived because trees use a lot of groundwater and would therefore aggravate existing water shortages. They suggest that grasses and shrubs are better crops to grow in these circumstances as these require less (ground)water. Other initiatives being utilized by the Chinese government include using silver iodide to 'seed' clouds and produce rain. This has had mixed results, with snow being produced instead of rain on one occasion in Beijing. Huge questions remain about the appropriateness of using such chemicals in heavily populated areas or the damage that might be inflicted upon fragile ecosystems by their repeated use. Although these outcomes might be unintended consequences of these practices, the possible impact of potential hazards should not be ignored. Social workers could undertake research to establish hazard benchmarks and assess the impact of disasters on communities holistically.

Conclusions

Environmental degradation has considerable implications for feeding the Earth's population and for initiating mass movements of people in search of food and safety. Neither the problem nor its solutions are simple. Population growth can outstrip food supply, whether for refugees in relief camps that stretch to accommodate all who come; those emigrating to better lands overseas; or those moving from rural to urban areas, where they feed the growth of megacities, as has occurred regularly for over fifty years, so that they have become termed 'Complicated places on the edge'.

A range of 'natural' disasters that occur with regular frequencies across the world raises questions about how limited professional resources can be extended to provide regular, reliable and skilled workers to assist in emergencies. An NCBS system could provide one effective answer. Social workers are found in all disaster situations, assisting people in obtaining water, food, clothing and shelter; seeking the peaceful resolution of conflict; helping in long-term reconstruction endeavours; or lobbying governments and multinational corporations to desist from exploiting either peoples or the land. This commitment to integrating considerations of the physical environment, the biosphere and human needs is a central component of green social work.

7

Environmental Degradation, Natural Disasters and Marginalization

Introduction

The UN estimated that natural disasters affect about 300 million people each year. These numbers can be swollen, and the devastation caused can be exacerbated, by the impact of human activities. The frequency and intensity of disasters seems to have increased over the past twenty years (UNEP, 2009). Yet, humanitarian aid fell by 11 per cent to US$15.1 billion between 2008 and 2009, most of the decline representing dwindling amounts provided by governments (GHA, 2010). People's capacity to respond effectively has also diminished. According to Webster et al. (2009) and Walker (2011), humanitarian relief lasting longer than three years now consumes 50 per cent of aid, compared to 35 per cent ten years ago. And two-thirds of this 50 per cent has been used to support people for eight years or longer. The refugee populations of Palestine and Cyprus feature prominently among this group. Although these two cases reflect the failure of politicians to find solutions to highly contentious issues, this has serious implications for humanitarian aid and questions its relevance in the long term. This includes whether its presence is simply compounding matters and letting politicians off the hook in the knowledge that minimum human needs are being met (Hoogvelt, 2007).

Dubious outcomes raise serious questions about the appropriateness and sustainability of aid endeavours and whether this is an effective way of intervening in disasters. The answers to these concerns are complex and contextualized. They involve considering what purposes aid is being devoted to as well as how it is delivered on the ground and to whom. Merging humanitarian aid with military purposes, especially those linked to terrorism as a result of George W. Bush's 'War on Terror', and the significant amounts of aid being diverted to Afghanistan and Iraq complicate this picture enormously (Hoogvelt, 2007). These issues also indicate that the distinction between 'natural' and (hu)man-made disasters is becoming increasingly blurred, often to the detriment of the aid recipients who become caught up in a politicized situation that is not of their making but who may be denied urgently needed emergency assistance, as seems to be happening in Somalia currently.

Preparing and planning for disasters can substantially reduce risk and damage caused to people, their livelihoods and the environment. Relief agencies have suggested that every US$1 spent on disaster preparedness measures can reduce post-disaster reconstruction costs by US$4 to US$7. Taking action now is essential because environmental degradation brought about by industrialization, urbanization and demands of growing numbers of people on Planet Earth have resulted in failures in the built environment and infrastructures that sustain them, to the detriment of peoples' livelihoods and well-being when 'natural' disasters strike. The impact of Hurricane Katrina on New Orleans is a recent example of the havoc that can occur when relief and government responses are inadequate, even when these involve the richest country in the world. The inappropriate responses to people of African American origins, older people and low-income families who were separated from their children during the relief effort after the levees broke in New Orleans demonstrate how those marginalized by society suffer most in such situations. This also highlights the politics of disasters by revealing how social, economic and political factors play a key role in transforming 'natural' disasters into (hu)man-made ones (Klinenberg, 2002). In short, the social positionality of those affected has a huge bearing on post-disaster outcomes, i.e. their chances of survival, potential to enhance resilience, and capacity to rebuild their lives to a standard of living the same as, or better than, that pertaining previously (Sharkey, 2007).

The multiple hazards of an earthquake, tsunami and nuclear reactor damage that occurred in Japan in early 2011 indicate how even the country thought to be the best-prepared nation for natural disasters could not cope with the demands for food, shelter, energy and care when large-scale disaster hits. Damage inflicted upon natural environments and infrastructures raises questions about how to develop technologies that meet people's needs without destroying the Earth. And it highlights the urgency of holding companies accountable for their decisions, as these affect all strata of society including the government, and not necessarily in beneficial ways. The floods in Pakistan during the summer of 2010 and earthquake in Haiti in the same year, with the subsequent cholera outbreak, illustrate the inadequacy of the speed employed in rebuilding infrastructures including sanitation and clean water supplies, and the failure of global responses to meet the needs of those affected by huge major natural disasters in low-income countries. These developments challenge social workers committed to eliminating marginalization and advocating for people's well-being to rethink their endeavours in such circumstances.

In this chapter, I examine social workers' involvements in these events, their critiques of traditional interventions and suggestions for providing more appropriate ones in the future. I argue that social workers' involvement has to be more preventative in its focus for practitioners to enhance the resilience of peoples and communities in minimizing danger, responding to catastrophes and rebuilding their lives afterwards. Maintaining a balance between protecting the environment, not making excessive demands on its resources, and meeting people's needs for decent standards of living is crucial to this endeavour, albeit difficult to achieve. I also consider how communities may become more effectively involved in co-producing the knowledge and skills that will create sustainable community development initiatives that prepare them for disasters, build resilience in its aftermath and enable those most affected at the grassroots level to feel they own the solutions developed. And I endeavour to examine how interdisciplinary teams of experts can work closely with local people to prepare themselves better for potential disasters, mitigate future calamities and build robust resilience that creates reciprocated relationships of caring for and being cared by others, between peoples and the physical environment.

Exploring Marginalization and Social Exclusion

Disasters impact with greater ferocity and damage on poor and excluded communities. This makes social exclusion an important concept for social workers to understand and integrate into their responses if they are to produce more sustainable and inclusive interventions. There are many definitions for social exclusion. The European Union (EC, 2004) popularized this term and defined it as:

> a process whereby certain individuals are pushed to the edge of society
> and prevented from participating fully by virtue of their poverty, or

lack of basic competencies and lifelong learning opportunities, or as a result of discrimination. This distances them from jobs, income and education and training opportunities, as well as social and community networks and activities. They have little access to power and decision-making bodies and thus often feel powerless and unable to take control over the decisions that affect their day to day lives.

The European Union's definition links social exclusion to marginalization. As a feature of social exclusion, marginalization emphasizes a disenfranchisement based on structural barriers, such as poverty, that prevent people from exercising their right to participate in society's decision-making structures and resource allocation systems. Socially excluded individuals and groups are stigmatized for their situation and often blamed for it. Pathologizing poor people for their plight has a dishonourable tradition in the caring professions, and was embedded in the way in which 'the social' sphere was conceptualized in the nineteenth century to produce 'deserving' and 'undeserving' claimants (Dominelli, 2004). It was and continues to be used to discriminate on a class basis. It is intriguing that in the UK, despite long-standing legislation on equalities, class, not specifically identified as an attribute of discrimination that is outlawed. Yet, class discrimination is central in the lives of working-class people, especially those without paid employment. Class, invisible in much social policy, acts as a taboo word for many, and is exposed in the differentiated impact of the recession and public expenditure cuts on poor unemployed working-class people, black and white. David Cameron expressed this beautifully when he claimed that 'We're all in it [the fiscal crisis] together', when this clearly is not so. Only poor people worry about how they will manage to feed, clothe and house their children and pay the bills for basic necessities like heating, lighting and water. Inequalities between different social groups are rising, not decreasing, between countries and within them (World Bank, 2011).

Byrne (2005) considers that definitions of social exclusion like those of the EU are embedded in a possessive individualism that undermines solidarity between people and reinforces social exclusion in and through social policies aimed at combating it. This insight is significant in disaster responses because poor, marginalized and excluded communities cannot afford individual solutions in mitigating risk. Examining how they can help each other individually through collective action as a community becomes an important means of overcoming this barrier. For that to occur, solidarity, mutuality, interdependence and equality, values that underpin anti-oppressive social work practice, can become useful tools in mobilizing people (Dominelli, 2004). Working along these lines means that green social work is about social work practice at its best: inclusive, holistic, egalitarian and transformative, caring for and being cared by people and the environment.

Those tackling social exclusion and marginalization question the model of economic development currently dominating the globalized world economy, namely that of a neoliberal capitalism that prioritizes the profit-generating opportunities of the few over the fulfilment of the basic needs of the many. One possible response to the concerns that this creates is that of a human and social development that requires economic decisions to be oriented towards placing social well-being over profit-making, and care for Planet Earth over its exploitation and pillage. This call for a socio-environmental economic agenda is not a new one. Back in the 1990s, Walker (1990) and Dominelli (1997) rejected the subordination of social policy considerations to economic, profit-making ones and called for the reaffirmation of the social sphere and inclusion of marginalized peoples in determining how to meet human need. Today, a socio-environmental emphasis suggests that instead of individual patterns of consumption determining the type of economy that societies should develop, a more holistic appreciation of collective needs should become its lynchpin. In this, individual needs would not be forgotten or ignored, but they would be satisfied in the context of ensuring that all of humanity's basic needs are met and that Planet Earth is also sustained as a living entity, with limits placed on the indifference towards the consequences of human action upon its beneficence that has featured in the relationship between current forms of industrialization and the world's natural resources.

Developments in various countries have sought to implement these ideas in practice through the formation of co-operative and social enterprises, e.g., Social Enterprise Africa. Other concerned individuals like Mohamed Yunus attempted to create income-generating opportunities for marginalized people excluded from financial services because they were poor, by creating the Grameen Bank. Credit unions and social enterprises sought to get poor people to pool their resources for the development of their communities and the greater good over a century earlier. For example, Desjardins in Canada grew from a small entity based in the province of Quebec to become the fifth-largest financial institution in the country by remaining true to the values of community development and human well-being (Shragge and Fontane, 2000). And it invests in ecological projects that protect the physical environment.

Community development is a crucial part of the social work repertoire, often lost under the hegemonic sway of clinical social work, especially in the United States, or under the managerially driven bureaucratic forms of social work prevalent in England. Community development is defined as the practices of individuals and groups seeking to bring about change in their communities by influencing society's wider institutions and power structures and mobilizing the skills and resources necessary for doing so (Craig and Mayo, 1995). Community development initiatives have a long-standing association with economic development, especially that linked to furthering capitalist modes of production (Dominelli, 2006) and promoting social capital and other networks that foster community empowerment and enable individuals to make fewer demands on government (Putnam, 2000). In contemporary society, community workers engage with these issues and others. A recent consideration in the social exclusion literature is that which arises from lack of knowledge of and connection with the digital economy, as this is creating other categories of marginalized groups (Wielm, 2004), giving community groups another issue to address.

Contemporary community development has begun to focus on environmental considerations and argued for social and community workers to engage with the physical ecology of the planet when working to improve jobs, local infrastructures and human well-being. I have called this 'green social work'. However, the term 'green' should not be associated with 'green' politics as espoused by Green political parties in the West. Whilst these are concerned about reducing consumerism, they do so in a way that ignores issues central to those that I focus on, namely, the lack of well-being among poor people in both the West and the Global South. And they ignore the significance of the market, particularly its current globalized form as articulated in neoliberalism, and the role of power relations in shaping decisions to which poor communities are subjected. For example, Catney and Doyle (2011) have highlighted how those engaged in climate change debates have ignored the needs of the majority of Earth's inhabitants who live in the Global South and are excluded from the market. Green social work places these people at the heart of its theory and practice.

A critique of the market is central to the enactment of 'green social work'. I adopt this stance in this book because the market increases inequalities between and among people by creating players and non-players. Consequently, those unable to access the financial resources necessary to become actors within the market-place are pathologized, excluded, disenfranchised and marginalized. Also, prioritizing the market and assuming that it produces a rational allocation of resources and services among people stigmatizes the non-players whose lack of access to material goods, services and power is deemed a personal pathology embedded in the personalities of those affected, rather than the structural inequalities inherent in the neoliberal approach to social relations (Culpitt, 1992).

Social workers' interventions in situations of environmental degrada-tion tend to be reactive. I argue that to conduct green community devel-opment, practitioners have to become proactive. And I suggest that preventative strategies that reconcile peoples' needs with those of the environment and its capacity to sustain itself have to be developed in full collaboration with poor and marginalized community groups. Realizing such plans will not be easy.

Social workers as community workers can mobilize people and encourage them to engage with community development processes. This means helping local residents to: engage in collective problem-solving endeavours from which they will all benefit; own the issues to be addressed; develop a vision and strategic agenda that their community owns; create effective local organizations; promote local leadership; iden-tify community assets, strengths and weaknesses; explore the nature of the social relationships between different groups constituting their com-munity; value the contributions of all community members and act inclusively; gauge how power relations are enacted in their community, who makes decisions on their behalf and how they can shape or influence these; create a community profile that shows the physical, economic, institutional, cultural and personal resources that there are in the com-munity; work together with other residents to prioritize initiatives and plan actions that include determining how to evaluate holistically the feasibility of proposed activities and effective use of resources; activate plans and constantly evaluate their progress; form alliances with other communities for specific purposes; access the expertise they need from other disciplines to ensure that they have the latest information at their disposal; and co-produce new knowledge for action in egalitarian part-nerships that support their objectives of making sustainable life-enhanc-ing decisions. A process of constant evaluation and reflection upon action to be taken will ensure that caring for people and the environment is an integral part of community development processes and included in every decision taken. Assessing and enhancing resilience is crucial to commu-nity enterprises aiming to create resilient communities are better able to manage and control change. The capacity to manage and control change according to one's wishes is a feature of robust resilience.

To become effective community development practitioners, social workers will require: cultural understandings and insights; appreciation of people's cultural and social aspirations from a vast array of sources; knowledge about economics and political and socio-economic systems, and access to information about the biosphere and natural resources and the consequences of people's activities upon these. The relationship between people and the Earth's flora, fauna and material resources is an interactive one that is multidimensional, fluid and constantly changing.

To understand how to mitigate risk, therefore, social workers have to dialogue with physical scientists and engage with their expertise on calculating risk from the physical environment and how to contain it within sensible boundaries. Ideally, they will co-produce knowledge together by working with local residents and enjoy the creation of innovative ways of viewing the world and solving the problems encountered (Lane et al., 2011).

Sustainable Development

The Brundtland Commission defined sustainable development intergenerationally to encompass the needs of people living in the present and future. Sustainability, as 'a form of development that meets the needs of the present without compromising the ability of future generations to meet their own needs' (Brundtland, 1987), permitted environmental considerations to come to the fore. The impetus provided by Brundtland led to the declaration of Agenda 21 at the Earth Summit in Rio de Janeiro in 1992. This outlined the main framework for sustainable development at the international level. The UN has a Division on Sustainable Development with a Commission on Sustainable Development charged with promoting sustainability at the local, national and international levels. Agenda 21 emphasized the need for everyone to have access to the information they required for decision-making; integration of social, economic and environmental considerations in that decision-making; and the imperative of enabling people to participate in formal decision-making structures.

Indigenous people played a crucial role in challenging this definition of sustainable development. They did so on the grounds that it did not include culture as one of its key tenets, and so they and others who met at the World Public Meeting on Culture in Porto Alegre, Brazil, in 2002, proclaimed Agenda 21 Plus Culture as an alternative. Their proposal insisted that cultural diversity was as significant to sustainable development as biodiversity and suggested that it was framed within a context of human rights, intercultural dialogues, participatory democracy, sustainability and peace. In this way, debates about sustainability could be conducted within an intellectual, emotional, moral and spiritual environment that was holistic and inclusionary. Their vision was formally adopted by cities and local governments at a meeting in Barcelona that produced their declaration of Agenda 21 for Culture in 2004. Its inclusive and holistic approach is consistent with the tenets of green social work as outlined in this book.

These initiatives are welcome, although their implementation has proven problematic. The focus in both the UN's initial definition of sustainable development and that of Agenda 21 was on individual consumption, while disregarding production. There was no critique of the market and its failures, especially with regards to the environment, where its contribution to environmental degradation can be seen as central to a market regime that has no moral or ethical basis for considering values such as the equitable distribution of resources; inclusivity that encompasses all peoples; caring for and protecting the environment and the Earth's bounty. As long as profit remains the most compelling determinant of market-driven actions, cheapness in overall costs, not care of people or the environment which can add substantially to the direct costs of producing goods, will remain the major deciding factor. Free-market indifference to both people and environment has recently been attested by the story of the thirty-three Chilean miners who survived sixty-nine days of being trapped underground when the San José Mine in the Atacama Desert near Copiapo, Chile, caved in during the summer of 2010. This disaster was blamed on lack of investment and maintenance, neglect of legislative obligations relating to safety in the industry, and indifference to the health of miners (Franklin, 2011; Macqueen, 2011). Thus, it was a (hu)man-made, not a natural, disaster. It could have been avoided if a duty to care for workers and the environment in which they work, and the rules and regulations already in place for employers to temper their treatment of workers, had been adhered to. Challenging such disregard of both law and workers' well-being is an area in which social workers can support local communities seeking accountability from employers. A recurring problem that seems to arise when trying to hold firms to account following a disaster of any kind is that the firm is sold off. This action complicates matters by disrupting continuity among the different actors so that records become lost or unavailable, and key players move on.

An economistic analysis of the costs of neglect in the case of the Chilean miners would be incomplete as it would only focus on directly incurred production costs. However, if calculations included the full costs of the mine owner's failure to act on repeated warnings and comply with legislative requirements as well as the cost of the rescue package, ill-health of these miners in both the short term and long term, and suffering that they and their families endured and continue to endure, a more realistic and complete appraisal of the risks associated with doing little or nothing could be ascertained. According to media reports one year after this disaster, jubilation at the rescue of the thirty-three miners in 2010 turned into fruit being thrown at them during the anniversary celebrations in 2011 because protesters objected to the lawsuit that the miners were launching against the government and mine owners for failing to uphold safety standards in the industry.

Protesters accused the miners and their families of being 'greedy' for seeking compensation. They also maintained that the costs of the rescue, estimated at US\$11 million, should prevent the miners from making further claims against taxpayers (Vergara, 2011). This exemplifies how competing interests among different segments of the working class can undermine attempts to protect both the environment and workers' health. Without some strategy for upholding such obligations, workers and the environment will continue to be subjected to exploitation by the mining industry. There is a role for social workers in developing alliances across the competing interests vying for attention in such situations. Crucial to their response will be engaging government in holding private corporations accountable for their decisions not to invest and then selling the firm without the accumulated debt, thereby leaving taxpayers responsible for paying outstanding bills, and mobilizing poor people to lead the deliberations because they pay disproportionately more taxes.

Besides being involved in supporting the miners through stressful moments in their lives and providing long-term support for them and their families, social workers can assist them in raising consciousness about their plight as a result of the culpability of both industry and government. Moreover, social workers can look for solutions that triangulate concerns about and care for the environment, current workers and future generations of workers and seek a win-win solution among all stakeholders involved. Looking at the holistic costs of doing nothing, or keeping on with 'business as usual', would constitute part of their arguments. These practitioners would also have to be aware of and address counter-arguments against this position. As some observers have pointed out, for example, the rescue of the Chilean miners was made possible by the technological ingenuity of the private sector. In response, it is important to acknowledge that it was not only mine owners who paid for this technology to be placed at the disposal of the miners and their families, but the Chilean taxpayer. Its use was paid for out of the public purse. Furthermore, there may have been hidden subsidies to industry to develop this technology in the first instance. These examples illustrate the importance of having a holistic picture of situations and overview of the knowledge that is held by all stakeholders in a dispute in order to hold successful dialogues and ease tensions between them. Such interventions fall within a rational approach to problem-solving. Without logical and coherent arguments, it is difficult to get disputing parties to listen to each other's concerns, encourage dialogue between them and make progress in solving problems that they both encounter. Social workers can use their skills in seeking non-violent solutions to conflict to achieve these goals.

Coping with the Demands
of a Growing World Population

The issues of creating sustainable development, defining what constitutes it and how it can be achieved are contentious. Analysts such as Walker (2010) contend that, to develop sustainable approaches to the world's problems, policy-makers and researchers have to tackle three major issues: climate change, a globalizing economy and demographic growth. I concur with this analysis, but suggest that finding solutions will not be easy as all three elements are hotly disputed and little common ground exists among protagonists. Among these three factors, demographic change has proven to be particularly thorny as it arouses strong emotions and, for some, deeply held religious beliefs. The issue is not a new one. In the eighteenth century, the Reverend Malthus (1798) argued that population growth could exceed economic capacity and thereby restrain social progress. He maintained that increases in agricultural productivity had limits and, therefore, it would be impossible for it to rise constantly into the future. Contemporary debates about the limits of the Green Revolution echo this theory. For Malthus, unless human beings curbed their reproductive inclinations (through sexual abstinence as was deemed appropriate then), population size would exceed economic capacity and food prices would rise substantially to meet demand. Such an outcome would, he claimed, further disadvantage and pressurize poor people. To prevent this happening, population growth had to be checked. Malthus assumed that this would occur through wars and widespread diseases wrought by nature that would reduce human numbers to reach an equilibrium that was sustainable in the long run. While he encouraged people to control their reproductive urges, in the absence of technological means to assist them in achieving this goal, he advised later marriage and sexual abstinence. I do not endorse Malthus' pessimistic view of society's capacities or the simplistic assumptions behind his theories, but the relationship between economic growth, the Earth's resources and population size has to be taken seriously, before its limits are reached. This requires more extensive and holistic research. In the meantime, there is a moral obligation to ensure that existing resources (in all their guises) are equitably shared so that no one goes hungry and every person and the environment are cared for.

There is nothing pre-determined about the limits to growth. Some scientists maintain that the Earth can sustain more people than are currently living on it, at least with regards to food production. They also suggest that population growth cannot continue indefinitely unless people as a whole, especially those living in greater wealth, refrain from consuming resources at the current rate (Guzmán et al., 2009). Others argue that population numbers will plateau out at about 9 billion and fall. The status of women is envisaged as central to achieving this outcome, and it is presumed that progress towards this goal can proceed more rapidly if women become better educated, involved in public life and more economically active (UN, 2003). This pattern has been observed in the West, where the population is not reproducing itself without immigration bringing in additional people (Guzmán et al., 2009). I am sceptical about such predictions, because there are too many unknowns in the equation for such statements to be made with any degree of certainty. People's behaviours change and the demographics may alter. While history can offer insights to the problems we face today, it cannot predict the future.

The UK provides evidence for this position. The population in the UK has not been reproducing itself internally for some time, as women either decided not to give birth or deferred it until much later. However, recent data in this country indicate that population numbers are growing internally for the first time in several decades, not through immigration but through natural growth among the British population as births outstripped deaths, and family size grew, from 1.64 children formerly to nearly 2 (ONS, 2011). Thus, while the issue of balancing population numbers is an important one, it should not be assumed that simply involving women in paid employment or higher education will automatically drive down population size. Specific people determine birth rates by making their own decisions in particular contexts, including socio-cultural and religious ones. At the same time, it is crucial for the equality of women throughout the world that women are encouraged to become better educated, participate in the workforce and public life, and have access to contraceptive devices so that they are empowered in making reproductive decisions for themselves, in consultation with those who matter to them. The role of social policy is to make these opportunities available to them. Social workers can support women in realizing self-determined goals and expanding their horizons about what they want to achieve in all aspects of their lives. They can also lobby for more enabling social policies.

'Natural' Disasters are Aggravated by Human Actions

The binary conceptualization of disasters as either 'natural' and attributable to nature or (hu)man-made and the responsibility of people is problematic. 'Natural' disasters are increasingly not thought to be only the products of Mother Nature. What might seem to be such an event, e.g., an earthquake in the middle of the ocean, may soon turn out to be a combination of both, as occurred in the 2004 Tsunami. The deleterious effects of a 'natural' event are aggravated by (hu)man actions that fail to respect the environment; take preventative measures that affirm safety in all matters; and eliminate poverty and social exclusion. Poverty in all its forms is prevalent among the most marginalized people in societies affected by disasters and makes their situations considerably worse. The links between (hu)man (in)action and poverty in disaster situations is evident in the different scenarios that are considered below. The stories that they contain demonstrate that there is much that can be done to mitigate risk and enhance resilience for all members of a community. Achieving the goal of robust resilience will involve collective action and the affirmation of an equitable sharing of socio-economic resources and the Earth's bounty.

Hurricane Katrina

The storm surges that breached the levees in New Orleans during Hurricane Katrina exposed the greater risk to marginalized people and highlighted how 'race', gender, age and class were either protective factors (for a member of the dominant or wealthier groups) or barriers (for those in the poorer subordinate group) in developing resilient responses. As the utilities and services infrastructures ground to a halt, those without personal means became dependent on emergency services, government or philanthropic relief for rescue and support. Curtis et al. (2011: 32) state that 'Floodwaters were higher, building damage was greater, and there was more human suffering and need for rescue' as the 'percentage of African Americans in a census tract rises'. Poor African Americans living in the 9th Ward, where poverty, 'race' and class came together to intensify marginalization and social exclusion, were the most severely affected. So, African Americans suffered more damage during the storm. Additionally, the help that they received afterwards was patchy and inadequate. In some situations it was inappropriate. For example, children were separated from parents when there was insufficient room in the transportation and shelter facilities, including the Convention Center intended to keep families together (Pyles, 2007). As a result, Sharkey (2007) argues, Katrina became a 'metaphor' for the deep inequalities that pervade America.

Curtis et al. (2011) suggest that, on the basis of measuring mortalities, missing individuals and flood height, older people – and, within that category, those of African American descent – had the highest proportion of deaths. Gender played a significant role. African American men were over-represented in the fatalities (Sharkey, 2007). And women, especially women heading households with children, fared badly on mental health assessments and other indicators of poor health (Greenough, 2008). Pregnancy also compounded their difficulties. Children were more likely to suffer disproportionately from disease, particularly infectious illnesses, because they have less well-developed immune systems and suffer from malnutrition. Children and women are also less likely to be consulted when decisions are being taken about disaster mitigation and interventions (World Vision, 2009).

Moreover, the lack of deeds to prove land ownership beyond doubt jeopardized African Americans' ability to claim assistance offered to owner occupiers to repair their homes. A combination of these factors increased vulnerability among African Americans and after the storm was over, they found that they were without homes and other important 'connecting' institutions such as churches (Curtis et al., 2011). Thus, their stock of both physical and social capital dropped when they needed it most and their recovery rate was slower than that of their white counterparts. Social workers have been involved in resettling people, reuniting families and helping people rebuild their communities. They were hindered in doing their job by the lack of resources and prior planning. The shambolic leadership offered by the federal government and FEMA (Federal Emergency Management Agency) and their inadequate responses highlight the importance of social worker involvement in building community resilience before calamity strikes and of having a number of well-trained volunteers on the ground, such as those that could be provided by a national, community-based service (NCBS), ready to go into action the moment disaster occurs.

Floods in Pakistan

The floods triggered by heavy monsoons in Pakistan during 2010 affected 20 million inhabitants, and devastated homes, power lines, road systems, communications networks and health care services. Buildings, factories, houses and 5,000 schools were destroyed, leaving millions of people homeless or with inadequate shelter. According to the International Red Cross, mines and artillery shells, washed down from Waziristan and Kashmir, posed additional hazards that can persist for considerable periods of time. Their presence could increase the numbers of people who might become disabled if the mines are set off for whatever reason. Flooding submerged one-fifth of the country's landmass, including important agricultural and food production areas. Crops and livestock were seriously affected, and ensuing shortages led to increases in food prices for already disadvantaged populations. The lack of sanitation infrastructures and clean water for drinking exacerbated vulnerability to diseases such as cholera, diarrhoea, gastroenteritis and malaria. The International Labour Organization claimed that 5.3 million jobs were lost through the floods, thus further aggravating poverty among those affected and driving down the country's economic performance. The relief provided was patchy. For example, the Ahmadi community, a religious grouping whose claim to be Muslim was rejected by the Pakistani government in 1974, complained that they were deliberately denied aid. Sikh minorities also protested that their needs for assistance were ignored in significant pockets of the country. Some Hindu groups were served beef in some relief camps in direct contravention of their religious requirements. Corruption by wealthy, feudal landlords who used relief funds for their own ends was also reported. Rich people were cushioned from the full impact of the floods by the resources that they owned before the floods, e.g., having cars that they could escape in.

The National Disaster Management Authority in Pakistan asked for external assistance, although the Taliban objected to such requests. The UN sought US$460 million for immediate relief purposes. Donations soon mounted to reach US$1,792 million. Saudi Arabia, Turkey and Kuwait were significant donors from the Islamic world, alongside the USA (the largest single donor) and the EU. The damage caused by the flooding is currently estimated to have cost US$10 billion.

Action Aid, one of the major relief agencies involved in rebuilding people's lives after this disaster, prioritized helping women and provided assistance to 234,000 people by the summer of 2011. Some of its initiatives involved cash-for-work schemes that paid people to clear debris and rebuild houses and public buildings. Other endeavours provided centres that created safe spaces for women and children, while some set up shops to enable people to earn a living. Working alongside government departments, practitioners in civil society organizations, including social workers, sought to ensure that people became better prepared for disasters. They helped local communities to develop contingency plans and identify focal individuals who could provide information and assistance on the spot in the event of future disasters. They also worked to ensure that all victim-survivors' names were entered in government records to ensure that they received compensation and other support that they might be entitled to in future.

Case study

Humanitarian aid is conducted largely under the auspices of NGOs. There are a number of high-profile ones that operate on the international scene – e.g., the Red Cross, Red Crescent Society, Oxfam, USAID, World Vision, CARITAS and others that are either very local or national and that are not normally known outside of their localities or country. These, like Action Aid and other NGOs involved in aid work, define themselves as relief organizations. Yet, they are performing social work tasks. Most relief workers do not possess specific qualifications in relief work, although most receive in-house training which may or may not be linked to a degree provided by a local university. The lack of formally recognized training for this group of workers is an issue for humanitarian aid organizations. While there is a need for appropriate training and checking of the workforce to establish that it is fit for purpose, behaves ethically, acknowledges culturally specific customs and issues, and will not take advantage of vulnerable people, the victim-survivors of disaster situations, a question arises about the relevant discipline that should assume responsibility for this training. Given that aid workers perform social work tasks, a social work qualification that emphasizes social and community development in disaster settings would be helpful. This would enable them to acquire a wide range of skills that would cater for individual and collective needs, and to work across a range of different settings, act ethically and behave in culturally relevant and empowering ways with both people and the environment.

The ALNAP (Active Learning Network for Accountability and Performance in Humanitarian Action) has recognized this as an issue and announced that a jointly convened conference between CERAH (Centre for Education and Research in Humanitarian Aid, based in Geneva) and ELRHA (Enhancing Learning and Research for Humanitarian Assistance), a network operating out of the UK, debated the issue in late 2011. These organizations acknowledge that there is a humanitarian studies curriculum and aim to consider 'the scaling-up [of] training and education of national humanitarian workforces and in high risk countries' (www.alnap.org/event/198.aspx). CERAH also offers a Masters and Certificate programme in humanitarian assistance, and so has a wealth of experience including various links to universities to draw upon.

Whilst I share the goals of this conference, I am concerned about the lack of a formal educational framework linked to a recognized career structure for humanitarian aid workers. As a social work educator, I would favour its becoming a field of expertise within the social work profession rather than being another example of its fragmentation. There are advantages in this. Social work is a profession recognized globally, with formal educational frameworks and career structures that vary from country to country to ensure that education and training is locality-specific and culturally relevant. It also has ethical frameworks and guidelines that can be used to prevent malpractice and apply sanctions if it occurs. It also has the breadth and vision of holistic, empowering approaches to practice including green social work that links personal needs to structural relationships and enables residents to determine what will happen in their communities.

The 2010 earthquake in Haiti

Haiti's 2010 earthquake was the worst in over 200 years and left 250,000 people dead and 300,000 injured. One million people were left homeless. The damage virtually flattened the capital, Port-au-Prince, and devastated government buildings, schools, hospitals, homes and other buildings. Cholera broke out a few months later and exacerbated conditions for an already suffering population. Poverty in the country aggravated the 'natural' disaster, as did corruption and ineffective governance structures. Haiti was rated 140 out of 182 countries on the income scale in 2009, showing that high levels of poverty existed before the earthquake. In the past year, its ranking has dropped to 145, and could decline even further (UNDP, 2009, 2011). In Haiti, 54 per cent of the population lives on less than US$1 per day, and the yearly per capita income is US$650.

The capital city, designed for 250,000, was home to 3 million people when the earthquake struck. Many of them were living in overcrowded slums and easy prey for unscrupulous landlords. Despite the country's poverty, the government's reconstruction efforts were hindered by the considerable sums paid as debt repayments to various countries, including the French, Venezuelan and American governments, the Inter-American Development Bank and the International Monetary Fund (IMF). Haiti's history of debt began as compensation payments to France for damages caused during its liberation struggle over 100 years ago. Some of the debt accumulated since was to pay for projects initiated during the Duvalier dictatorship years, but which did not benefit the general population (Levy, 2010). Some repayments were cancelled prior to the earthquake, under pressure from activist groups such as Jubilee 2000. However, a sizeable sum remains to be paid as not all debt incurred before the 2010 earthquake was cancelled. Since then, the IMF has provided Haiti with further loans, not grants, although it has promised to revisit this issue once the rebuilding linked to reconstructing the country, after the earthquake recovery work, is completed (Levy, 2010).

Case study

Children are a particularly vulnerable group during disaster situations. Haiti, like Sri Lanka during the 2004 tsunami, had a large number of children whose parents and/or relatives were temporarily missing and had to be found, and there were those whose parents had died in the earthquake and needed formal arrangements made for their care. In the ensuing chaos, undesirable events occurred in these children's lives when some adults preyed on them and abducted them for several purposes, including: prostitution; trafficking; sexual abuse; and adoption by childless people overseas with little or no connection to the children's country or culture. In Haiti, the tradition of the 'restavec' which involved parents fostering children with people known to them (including relatives) was abused by some individuals who claimed they were taking children abroad for a temporary period to give them a better start in life.

Social workers are well placed to protect children in such circumstances. However, this does not always happen immediately because insufficient numbers of practitioners are available to cater for all the children, the necessary local facilities including schools may have been destroyed by the disaster, and social workers themselves may have been victims of the disaster. In such situations, UNICEF is particularly active in responding appropriately and quickly finding people who can help including bringing resources from across the world. Once in place, these workers cooperate with local professionals, especially social workers and health professionals to reunite children with their families and settle them into normal routines like schooling.

The Sri Lankan tsunami and Haitian earthquake exposed similar problems around the lack of social work capacity, including in community development at the local level. This issue must be addressed speedily by governments, who should provide local academics with the resources necessary for meeting this need, and build on the reservoir of goodwill from overseas, including that available through the International Association of Schools of Social Work (IASSW), to develop educational capacity and get resilient systems in place before the next disaster strikes. Seismologists tell us that there will be another big earthquake in Haiti before long, although they cannot put a specific time on it. Acting now will result in more resilient local and professional responses in future. Capacity building in education is an important dimension of this work. It will also ensure that there are skilled professionals on the ground ready to engage children in the appropriate manner and safeguard their well-being, including reuniting them with families or kin and ensuring that they participate in normal routines like attending school. They can also listen to children's suggestions about disaster prevention and preparedness and help take these forward.

Social work academics from several universities in North America have been working on developing educational infrastructures for social workers in Haiti for a while. For instance, a long-standing one has been supported by the University of Montreal in Quebec, Canada, which has a sizeable Haitian-origined population. The location of these initiatives can be important given that they can draw on Haitian orgined people living overseas to work directly with colleagues in Haiti because they speak local languages and are aware of local cultures and traditions. Following the 2010 earthquake, academics at the University of Montreal joined up with others, including those at Durham University and the University of Barcelona, and in civil society organizations in Montreal, London and Madrid, to link up with social work academics and practitioners in Haiti to develop capacity in social work education there. Additionally, the IASSW encouraged members in the Schools of Social Work in the Caribbean to support Haitian colleagues. This they did by going to Haiti in the immediate aftermath of the disaster. However, all these endeavours were hindered, and continue to be, by the lack of capacity and resources to fund long-term development work in the professions in Haiti. The limitations to 'piggybacking' on existing systems and support from governments, NGOs already active in the field and Schools of Social Work linked to specific individuals became evident rapidly. The lack of resources to support Haitians effectively led to a loss of morale and some active players withdrew from the initiative because they felt undermined and unable to continue (personal communication).

Rebuilding Haiti has proved slow and, eighteen months after the earthquake, around 650,000 people were still living in tents in Port-au-Prince and having to deal with another season of hurricanes and tropical storms such as Emily. Poverty has hindered reconstruction efforts, alongside disputes about land ownership and other resources (Action Aid, 2011). However, there are centres where people can congregate to share experiences and celebrate their survival. These sites engender hopes for the future and enable people to begin to heal and deal with the traumas that they experienced as a result of the earthquake. Besides offering psycho-social support as stipulated in the Inter-Agency Standing Committee (IASC) guidelines, and rebuilding communities, social workers in international civil society organizations can monitor the use of donated funds and ensure transparency in their delivery, i.e. that funds reach intended recipients.

Action Aid (2011) has identified the dangers of not building permanent homes for the people of Haiti, as slums are created when temporary shelters deteriorate and have to be replaced at significant additional cost. Action Aid also highlighted the predations perpetrated upon local people by slum landlords and criminal gangs, and masses of unemployed people living in socially marginal areas. Such living conditions can exacerbate the ill-health of individuals and communities attempting to rebuild after surviving significant traumas. Failure to provide resources and action that ameliorates these situations can produce social unrest. A housing strategy and provision of long-term employment opportunities would be essential in avoiding further catastrophic collapse in such marginalized social and physical environments. Social workers active in the community can assist in creating the relevant housing action plans and monitoring developments.

Rebuilding housing in Haiti might require a replacement for the cadastre system inherited from former colonial power, France, which was never fit for purpose (Action Aid, 2011). Multiple claims for one plot of land, loss of deeds, or not having deeds available, despite families having occupied these lands for generations, complicate the rebuilding process. The Organization of American States (OAS) has offered to help Haiti with its land reform. To be successful, such change must involve poor people and address their needs while responding to existing land-owners' demands for compensation to enable them to cede claims to ownership. Haitians are keen to avoid the creation of useless developments typified by Duvalierville, a city 50 miles north of Port-au-Prince created by former dictator (Papa Doc) François Duvalier, who resettled middle-class people there. However, as jobs were not forthcoming, the city, with an unemployment rate of 80 per cent, has become a run-down, unattractive location.

In the meantime, the government has declared 450 hectares of land in the downtown area of Port-au-Prince necessary for the construction of public buildings destroyed in the earthquake. It will only pay compensation if people can produce deeds to the land. This will be difficult for some legitimate owners to do. If the government can include social housing and recreational spaces for urban landless people in such schemes, this could facilitate job creation and develop resilient communities for the future. Housing and land tenure issues were not included in the Haiti Post-Disaster Needs Assessment. Housing is not prioritized in the remit of the Recovery Commission either. The lack of focus on housing construction and engagement of local people in housing design and formation is poor development practice. Without the community's involvement, reconstruction plans may include costly mistakes. Membership of the Recovery Commission is composed of equal numbers of Haitian and non-Haitian elites, allegedly to ensure transparency in the proceedings. This arrangement does not augur well for the formation of strong, democratic institutions run by Haitians and working for Haitians to secure government accountability. Haiti's reconstruction plans should be sustainable and inclusive of the grassroots, rather than being handed down by those in the upper echelons of society.

Having local residents participate in these development processes is an important part of building robust resilience in communities. This helps them to acquire confidence in their skills and make decisions that enable them to control their lives better and hold politicians accountable. This includes their being able to engage in the effective deployment of resources, including land and its uses for food production, civic and recreational amenities, housing and industrial purposes. To engage people successfully in collective action, there has to be trust among the constituent groups and between them and their rulers. They also have to believe that they can: achieve their economic, social and cultural goals; build community capacity; formulate new knowledge based on recognition of the worth of the diversity of opinions and views held by all members of their community; and develop their full potential by working together. Through such cooperation, they can contribute more effectively to the nation-building process, a feature that is important in establishing the governance structures and infrastructures to develop the sustainable economy and job opportunities that will create the wealth and durable environment required by Haitians. Acquiring new knowledge and formulating innovative attitudes towards the challenges that face them can enable communities to redefine their problems and see fresh opportunities and possibilities in what might seem intractable problems, especially if they work alone within well-worn tramlines. Community development involves collaboration, local leadership, allies from both within and outside the community working with local people, and boundless energy.

Social workers can assist residents in implementing these tasks. Social work practitioners and educators have been involved in supporting capacity-building initiatives in Haiti before and after the earthquake. After the earthquake in 2010, the IASSW and the International Federation of Social Workers brought together academics and practitioners from different countries to facilitate capacity building in social work education and community reconstruction efforts.

Addressing social divisions: disability

Social divisions such as ethnicity, gender and age are important entities that differentiate the experiences of disasters. Disability is another of these features that requires specific attention during and after disasters. Disability figures tend to rise because limbs become particularly vulnerable. Disability rates rose in both Haiti and Sichuan after their earthquakes, for example. Speed in getting medical supplies to such localities is essential in preventing gangrene from spreading uncontrollably through injured limbs. In situations where medical assistance has been delayed for a considerable period, as occurred in Haiti, for instance, medical practitioners amputated limbs rather than allowing wounds to fester in order to save the person (personal communications). While this may have been the appropriate medical response in the circumstances, from a social work perspective it is crucial that resilience is promoted before disasters to find local solutions to potential hazards by drawing on local knowledge and resources, e.g., boiling sea water to bathe limbs constantly where it is easily available. Local knowledge exists in all disasters and can be utilized to minimize health risks caused by delays in receiving aid, to plan better for such events, and to find ways of assuring supplies and responses before disasters occur. The reactive nature of UN responses is, I think, a barrier to humanitarian aid. It needs to develop a more proactive state of readiness. We know that disasters are increasing in frequency, involving ever greater numbers of people and habitats, and so should be preparing to intervene more effectively and rapidly as a matter of urgency (Dominelli, in press).

Involving communities in solving these problems is crucial. They have local knowledge to hand. Raising consciousness about the risks entailed and helping local people build on existing strategies for survival are crucial to building on existing strengths and develop community and individual resilience and well-being. Zhu and Sim (2010) writes movingly about work undertaken with disabled children in schools after the Sichuan earthquake, as they were taught ballet to build self-confidence and robust resilience among young people and their teachers. Children can become ambassadors in such a process. Training children in disaster reduction techniques and what to do in the event of a natural disaster is a risk-reduction strategy that can save lives and property. Including such training in the classroom can assist such developments considerably. This strategy is being adopted and promoted through schools as a regular part of the curriculum in Bangladesh, China, Japan, El Salvador, the Philippines and other countries. Listening to children's voices can promote resilience initiatives and assist in developing risk-reduction strategies (World Vision, 2009). Building adequate shelters, ensuring these are placed in safe locations and ready for use if needed, and siting early warning systems where local people can hear them are central to preventative endeavours. The actual timing of a disaster may be difficult to predict, but people should be trained to react quickly. Multi-use buildings might be better investments than purpose-built shelters in poor communities where funds are scarce. Schools and community centres often act as such buildings, given that there are considerable periods when they would otherwise lie empty.

   Social workers have an important role to play in all of these elements of disaster preparedness. They can advocate for and coordinate action regarding the: capacity building; removal of debris; declaration of safe relocation sites; construction of safe, accessible and affordable permanent housing where local residents actually want to live; and participation of people, especially marginalized children, women and men, in planning processes that will enable sustainable developments to occur in an orderly, transparent manner.

Conclusions

The division between natural and (hu)man-made disasters is becoming increasingly insignificant. Boundaries between them blur because (hu)man activity has impacted upon the landscape and 'natural' terrain. Focusing on prevention, mitigating human impact, improving individual and community resilience and creating long-term sustainable solutions is the way forward. Social workers have a crucial role to play in facilitating the development of such plans by working alongside specific communities and promoting their well-being. However, social workers are hindered in this work by the appropriation of their skills by others, their failure to engage in both local and global issues, the lack of capacity and resources for long-term endeavours, and a limited role in UN disaster intervention committee structures.

8

Scarce Natural Resources and Inter-Country Conflict Resolution

Introduction

The Earth's natural resources such as land, water, energy supplies and minerals are being exhausted by the demands of agribusiness and industrialization processes that follow Western models of development, and increasing population growth. The spread across the world of the Western industrial model originally developed in the UK, a country with a small population, is clearly unsustainable for contemporary demands. It has been unable to sustain high standards of living even for the few who benefit from neoliberal capitalist development now, so it is unlikely to be capable of catering for the growing numbers of the world's population, which the UN predicts will exceed 9 billion by 2050. Whilst I do not endorse Malthusian gloom over this issue, unless ways of raising people out of poverty, and promoting sustainable development and healthy lifestyles for all of the world's peoples, are found, the future could be very bleak for current generations and those to come, and the Earth's flora and fauna. Sustainable development in the environmental context has a huge agenda. It aims to: empower people, especially marginalized groups including women and children; overcome poverty and hunger; maintain environmental sustainability; reduce the spread of disease including HIV/AIDS and malaria; tackle maternal ill-health and child mortality; improve primary education; and work in partnership with a range of stakeholders globally to enhance livelihoods and protect the ecosystem.

The growing realization that the Earth's natural resources are scarce has intensified competition and tensions between countries. The UN has sought international agreement to bring about an orderly resolution to such disputes. Although international bodies and many non-governmental organizations (NGOs) and civil society organizations at the international level agree that linking development, poverty eradication, livelihood protection and conserving biodiversity is essential to development, this may not be the approach taken at the national level where local policies are made. For example, Poverty Reduction Strategy Papers (PRSPs) do not focus on both poverty alleviation and environmental protection, including that of biodiversity (Pisupati, 2004). Agreements may be reached internationally while being ignored nationally and locally. While the Helsinki Rules were agreed internationally in 1966 and enshrined the principle of equality in the usage of water, they were not always adhered to as local agreements tend to favour existing usage and allocations (Anand, 2004). This is clearly demonstrated in the case of water shortages in Central Asia described below. Population movements exacerbate conflicts over resources through claims and counter-claims aimed at asserting sovereignty over disputed areas as people seek refuge when tensions erupt into violence to uphold claims over disputed resources.

In this chapter, I examine how social workers can support inter-country initiatives in resolving potential conflicts over scarce resources to achieve win-win *situations* for all, including Planet Earth. To this, I add a consideration of supportive policy initiatives at local, national, regional and international levels. For practitioners on the ground, this means engaging with local communities to widen access to scarce resources, including drinking water which is seen as a universal need, and sustainable use of water for domestic, agricultural and industrial purposes. Such interventions will involve social workers more fully in sustainable community and socio-economic developments at the local level. Within such interventions, governance practices become a central consideration in the discussions held for their co-development with local groups. Other scarce commodities that could engender conflict include land, energy and food. We have already seen food riots spread across the globe in 2008 when food prices rose substantially and exacerbated precariousness in the lives of poor people. Some analysts have argued that the invasion of Iraq in 2003 was to assure Western, especially American, supplies of oil. And Naomi Klein declared in the *Shock Doctrine* that disasters are used to shift resources towards those with money as they grab land and other resources away from vulnerable people to profit from disaster crises. Similar allegations were made following the 2004 Indian Ocean Tsunami. Social workers have a role in helping communities to assert their interests and ancestral claims to land and its resources against such predations and to protect their environment from being further degraded.

Scarce Resources and the Dynamics of Place and Space

Conflicts over resources occur not only by virtue of their scarcity value but also if exacerbated by population movements, especially of poor people. Such conflicts can escalate into violence and have even more devastating effects on the environment where bombing and mining areas become routine. Understanding population dynamics and their impact on the environment requires an analysis of the differentiation within a population: age (whether young or old); ethnicity; location (whether urban or rural; in the Global North or Global South); family size (number of children and adults); household composition (whether smaller households are becoming the norm) – alongside their numbers. Household size has been dropping in countries such as Brazil and China, where declining population fertility has been accompanied by a growth in single and dual income couple households without children who have different consumption patterns from family households with children, whether single-income earners or not. The households responsible for this new demographic trend have more wealth, consume more resources, and produce more greenhouse gas emissions (Jiang and Hardee, 2009). Similar issues arise elsewhere, and some countries have sought to mitigate these pressures in their cities to an extent. For example, Manizales, a city in Colombia, has endeavoured to reduce the impact of climate change caused by high-density populations settling on marginalized and vulnerable lands by making more appropriate plots available to low-income populations (Velásquez, 2005).

The links between poverty and risk, in predicting vulnerability to climate change scenarios, environmental degradation and conflicts that might ensue, are not straightforward (Balk et al., 2009). Maps that contain spatial and population data showing where sanitation, drainage and water supply structures are located can help identify sites of potential hazards for policy-makers, emergency planners and engineers seeking to minimize their effect. They may also pinpoint conflict hot-spots where disputes over scarce resources including built infrastructures and utilities have been simmering for some time. Such maps are being developed by the United Nations Population Division through the Global Rural–Urban Mapping Project (GRUMP), based in the Socioeconomic Data Applications Center at the Earth Institute, Columbia University, and researchers at the City University of New York and the Population Council in New York City.

The impact of climate change, environmental degradation and conflict situations on individual and group migration is uncertain, especially as there is insufficient empirical evidence to indicate what would happen in particular scenarios. A range of diverse factors have to be taken into account in calculating the effects of human movement. Their outcomes would depend on the location of extreme weather events, environmental degradation at such sites and populations affected. A specific example of the exacerbation of potential tensions between nomadic pastoralists and sedentary farming communities in conditions of scarce resources, where such tensions become aggravated by armed conflicts with deleterious consequences for local people, is Darfur (Bachir, 2008). Here, the impact of political marginalization, limited economic opportunities and environmental pressures was intensified by the breakdown of traditional social structures and local arrangements for mediating between disputing groups (Edwards, 2008). Rwanda provides another location where disputes over scarce resources (water) contributed to tensions that ended in the crime of genocide.

Other countries have sought to utilize non-violent means in resolving conflict over demands for natural resources. A current instance is the talks being held between Egypt and Ethiopia for the management of the waters of the Nile to meet the needs of growing populations in both countries and their desire to improve their standards of living for their disadvantaged peoples (Kameri-Mbote, 2007). Their approach contrasts with the violence that has already erupted over the control and use of the waters flowing through the Kaveri river.

While involving several other states, the main parties at the centre of the dispute over the Kaveri river are the states of Karnataka and Tamil Nadu in India (Anand, 2004). This controversy is of long standing and goes back to a settlement initially formulated in 1807 during Britain's colonial rule of India. It has been the subject of on-going clashes since, with serious violence erupting during 1991 and 2002. The courts have sought to resolve this dispute during the intervening years, but many have questioned their rulings to the present day. The riots during the 1991 conflict were so grave that schools in Bangalore were closed for a considerable period. The situation has been aggravated by increased need for water to cater for the growing demands of agriculture, industry and populations. Conflicts over water can escalate to affect other resources. In the 2002 conflict, a power transformer was blown up by pan-Tamil militants, and, by affecting energy output, this indicated their hostility to the failure of Karnataka authorities to supply water as stipulated by the courts. Moreover, attempts to settle this conflict involved the central government and Supreme Court of India mediating between the disputing parties (Anand, 2004). Social workers can play a key role in energizing dialogue between local community groups in such disputes. Person to person contacts can facilitate understanding, where it is lacking, more easily than formal bureaucratic approaches.

Social Workers' Involvement in Situations of Resource Scarcity

Conflicts over limited resources can emerge at any moment and in any country, because any type of scarce resource can be involved. Although these usually focus on land, minerals, air space, water, housing and employment opportunities, other resources can be involved. Below, I consider some crucial resource shortages involving housing, jobs and water that can lead to conflict. I also examine how social workers can support people in resolving their differences.

Shortages in housing and employment opportunities

Social workers have seldom been involved in policy discussions aimed at ensuring an equitable sharing of resources between countries. Yet, they are among the key professionals working at community level to reconcile differences when struggles over scarce resources erupt into violence. Housing has become a scarce commodity. Housing shortages can be aggravated by the lack of affordable housing, and low levels of housing construction. There are already 200,000 slums in the world that are inhabited by more than 1 billion people without access to low-priced housing. Social workers mediate between warring groups to find ways of moving forward by encouraging dialogue and non-violent conflict resolution strategies among them. One example of these at community level was the contribution of the School of Social Work at the University of KwaZulu-Natal, Durban, in bringing about an end to the fighting over scarce housing and jobs between members of the Inkatha Freedom Party (IFP) and the African National Congress (ANC) in a neighbourhood in the city. Disputes between these two groups had already involved violence before and after majority rule, including in 1994 when disputes in several cities, particularly Johannesburg and Durban, reflected wider controversies between them.

During the liberation struggle, supporters of the ANC were locked in armed conflict with the Zulus under Chief Mangosuthu Buthelezi, who led the IFP. The fighting between them persisted after the end of apartheid and the country becoming liberated in 1994 under the leadership of the ANC, with Nelson Mandela, its then head, becoming President. Even in local communities, violence was endemic as people sought decent jobs and housing in conditions where these were scarce. Staff and students at the School of Social Work commenced community development activities to enable these two groups to talk together and pool resources for the better good. They had a reasonable degree of success and the dialogue that ensued between these two previously antagonistic factions reduced tensions considerably.

A different set of clashes arose simultaneously in various locations in South Africa in 2008 when migrant workers, asylum seekers and refugees came under attack. Those from Zimbabwe and Mozambique were particularly at risk. There were 60 migrants killed and 100,000 displaced between 2008 and 2009 during these violent clashes. Social workers identified the importance of protecting undocumented migrant workers from attack and supported both migrant workers and South Africans in dealing with the traumas that this violence precipitated on both sides. The School of Social Work at the University of KwaZulu-Natal intervened in ethnic conflicts involving migrant workers, asylum seekers and refugees. It asked the government to resolve the matter using a social justice framework and to build on examples of good race relations that had featured in South Africa since 1994. In this case, the UN's Human Rights regional office in Pretoria also worked with the South African Human Rights Commission (SAHRC) and provided US$100,000 to initiate an anti-discrimination project that would be run locally by the SAHRC. This project aims to support individual victims of this violence, create a more supportive environment for migrants by raising awareness of anti-discriminatory positions and legislation, and develop capacity within SAHRC itself.

Scarce water resources

The Earth is 71 per cent water. However, only 2.5 per cent of it is fresh water and much of it is stored in the glaciers of Antarctica. Only 1 per cent is actually available for human use. However, it is unevenly distributed across the globe (De Moor and Calamai, 1997). Water poverty has now become a social science concern, much like fuel poverty. Although the concept of water poverty is not as well known, it has been defined as a family having to spend more than 3 per cent of its income on water

(Benzie et al., 2011). Of the fresh water available, 8 per cent is used for domestic consumption, 10 per cent for industrial purposes and 82 per cent for agriculture. Waste occurs in all areas of usage and is substantial. Much of it could be avoided. For example, its waste in agriculture is considerable, but wastage occurs primarily through inefficient irrigation systems. De Moor and Calamai (1997) argue that this waste could be greatly reduced by the removal of publicly funded water subsidies for agricultural purposes. Moreover, they argue that water subsidies for any sector of society, including domestic consumers, encourage waste by lowering the price of commodities and distorting market mechanisms. On top of that, they maintain that subsidies benefit rich people more than they do poor people who pay more per unit of water consumed, often because they are not connected to cheaper mains supplies, or because their consumption is metered. Developments in the USA, China, Central Asia and Bolivia, among other places, bear their theory out. Water pollution is another problem that could be substantially reduced, improving human health in the process, especially among children who suffer disproportionately from diarrhoea and other water-borne gastric diseases.

Drinking water is a scarce and unevenly distributed resource. Turning it into a commodity from which huge profits can accrue will exacerbate tensions between those who have water and those who do not. The Middle East, for example, is home to 5 per cent of the world's population, but holds only 1 per cent of its freshwater resources. To avoid water shortages, it has to seek fresh water through a variety of means, including trade, desalinating ocean water and locating deeper aquifers. Disputes involving water could become more significant in future (UNESCO, 2006). Water wars could erupt anywhere, given that about 1.1 billion people in the world are without potable water. Lack of drinking water is not the only source of conflict over water. The pollution of water sources by industrial firms, the shortage of water for commercial use including irrigation and industrial manufacturing processes, and water-based industries such as fish farming, can exacerbate disagreements about the ownership and use of water. And there is the issue of holding large users of water accountable for the efficient use of this scarce resource and maintenance of the purity of supplies for all users.

The United Nation's Hydrological Programme and the World Trade Organization are both charged with facilitating the resolution of disputes about water between member states. They are also involved in training professionals to raise awareness about the issues and facilitate peaceful conflict resolution strategies. This is a task which social workers are suitably qualified to fulfil. The location of water resources and their flowing through a number of different geographic regions or states highlights the importance of collaboration between countries and the recognition of the need to focus on interdependencies between them to devise equitable solutions that meet the needs of all. UNESCO has been working on producing a map of aquifers that cross national boundaries to enable neighbours to look for ways of sharing such resources. Conflicts over water can occur in any country. In the 1960s, the pollution of the Great Lakes straddling the Canadian and American border brought local communities out to demand that they be cleaned up so that the environment could support fish and other wildlife as it had before. The American Congress passed the Clean Water Act in 1972 that began the process of cleaning them up. Sometimes, the tensions can be domestic ones, as occurred in China where many cities and rural areas lacked safe drinking water. To address the issue, the Chinese government introduced the Water Pollution and Prevention Law. As a centralized authority, it can enforce the conditions contained in such legislation.

In 2000 in Bolivia, protests erupted in Cochabamba when a multinational consortium sought the ownership and management of the city's public water supply. This privatization was muddied in wider conditions imposed by the World Bank for authorizing further loans to the country. The World Bank was concerned that the regime was too corrupt to manage either the water resource or its privatization properly because, under its stewardship, poor people had to purchase water of unreliable quality at extortionate prices from those bringing it in trucks and handcarts. The Bolivian population in Cochabamba did not want its water privatized, and resisted the government's moves. Eventually, others joined the cause, and what began as protests over water became enlarged to include workers, peasants and middle-class professionals, including the police and teachers, striking for pay rises. Action around their joint grievances led to violent confrontations involving many areas of the country. This dispute illustrates how many other concerns can become triggered by conflicts about water.

Central Asia is also a site of disputes over water, particularly those involving the rivers Syr Darya and Amr Darya. The loss of habitat in this region amply demonstrates the devastation that can be caused to the environment by (hu)man intervention and is encapsulated in the events surrounding the desiccation of the Aral Sea, which has been shrinking in size since the 1960s. It has now been reduced to two saline lakes, thereby depleting fish stocks and impairing the health of people living within its catchment area. The damage was caused by the decision of the then Soviet authorities to divert the two main rivers flowing into the Aral Sea for agricultural purposes, namely the growing of cotton which requires huge quantities of water for irrigation. Cotton producers in Uzbekistan and Turkmenistan are major beneficiaries of this policy.

These water problems are complicated by border disputes between the five riparian countries (those through which the rivers flow) with a direct interest in the waters of this area, including that of the Syr Darya and Amr Darya rivers, namely Kyrgyzstan, Kazakhstan, Turkmenistan, Uzbekistan and Tajikistan. Disputes over water are complicated by: the Soviet legacy; different possible combinations of leasing natural resources, including water and energy, across the borders between these nations; political instability; questions about national sovereignty; corruption; poorly developed transportation infrastructures; ethnic and religious divisions between the countries; disputed borders; and poverty among significant proportions of their populations (ICG, 2002).

Allouche (2007) suggests that the water problem in Central Asia is its uneven distribution rather than a lack of supply, and that this is the result of Soviet decisions about water allocations that are inappropriate for a region that is now composed of five independent countries, all relying on the waters that flow through this territory. Thus, conflicts over water have become issues of governance and management of the resource rather than its availability. The situation is unstable, and conflict is possible at any moment. This is more likely to occur during times of drought. Although the five states involved agreed to abide by the Soviet agreements in place to harvest the cotton immediately after the demise of the USSR, they have been unable to agree on the way forward since. Under the Almaty Agreement of 1992, they agreed to keep to the Soviet allocations whereby the upstream countries of Kyrgyzstan and Tajikistan provided water to Uzbekistan, Turkmenistan and Kazakhstan. At the same time, the three located downstream (Uzbekistan, Turkmenistan and Kazakhstan) agreed to make energy resources available to the two countries upstream. Additionally, all five countries were to set aside 1 per cent of their GDP to restore the Aral Sea. However, this Agreement has become moribund, as each nation's politicians have become more interested in pursuing their individual interests.

The IFAS–ICWC, the key body established to resolve the water issues is made up of the International Fund for Saving the Aral Sea (IFAS) and the Interstate Commission for Water Coordination (ICWC). It has been unable to do this because divisions over whether to use water to produce hydroelectric power or to irrigate the fields have exacerbated conflicts over water between these countries and made them seem intractable. Kyrgyzstan is particularly keen to treat water as a commodity that can be traded on the open market. This approach is objected to by those irrigating cotton fields. Upstream countries simultaneously complain that the downstream ones that receive water do not pay their fair share of the funding needed to maintain the facilities that supply them with water. Moreover, they allege that downstream countries use the water received, especially its agricultural allocations, inefficiently. Experts have estimated that Uzbekistan utilizes four to six times the amount of water needed for irrigation purposes because the water is not metered and an economic price is not being charged for it (Allouche, 2007). Whilst not paying the economic price for water, both Uzbekistan and Kazakhstan are charging Kyrgyzstan market prices for energy supplied. Thus, in the arguments between them, the lack of water security becomes a way of pitting energy security against agricultural security.

As agreement about the use of water has not been secured, arguments about water ownership and usage involving Kyrgyzstan, Tajikistan, Uzbekistan, Turkmenistan and Kazakhstan have become increasingly complex as demands for water have risen among three key user groups: the populations of these countries; industry; and agriculture. Conflict among different ethnic groups claiming ownership of the water in the area's two major rivers, the Syr Darya and Amr Darya, has been a feature of this dispute (ICG, 2002). Matters became extremely serious in the summer of 2010 when violent conflict erupted on the borders between Kyrgyzstan and Uzbekistan and local politicians sought to protect national interests rather than calm ethnic tensions in the region. Consequently, violent clashes flared up in Osh: 200 people died and 400,000 had to flee. In the ensuing riots, claims and counter-claims about local status and positions flowed between the Krygyz and Uzbek peoples, with each accusing the other of being the dominant group. Eventually, local established residents ousted migrant workers to ensure that only the 'true' residents of the locality benefited from the area's resources (Stern, 2010).

Conflict refugees

Refugees were created by the systematic destruction of property that followed this violence as people sought safety. Social workers were asked to assist the Uzbek refugees who crossed national borders into Uzbekistan. The conflict in Osh has deep roots complicated by ethnicity, with Uzbeks living there feeling insecure despite being more economically advantaged than the Kyrgyz, and the Kyrgyz believing they were second-class citizens in their own land, despite the record of cooperation between them. The refugees escaping Osh wanted shelter and peace. They had not initiated the attacks, but were caught up in violence and fled for their lives when their homes were burnt and people killed for being different from the dominant ethnic group.

Refugee camps are supposed to be places of safety. This status cannot be guaranteed and they can become danger zones when violence increases.

Such violence is often aimed at women and children, especially young girls. For example, in some emergency camps created to assist people after the 2004 Indian Ocean Tsunami, adolescent girls were particularly vulnerable to sexual harassment and abuse (Fisher, 2005). Similar issues have been raised in the Dadaab camp.

Although at greater risk, children are remarkably resilient and have demonstrated their ability to cope with traumatic situations, provided that they: have active support from adults; are purposively engaged in solving the problems that they face; receive support from their peers (Boyden and Mann, 2005); and have access to adults whom they trust and who are interested in them (Werner and Smith, 1992). Children's rights and child-centredness become important issues that social workers can explore in disaster situations with parents and community leaders to facilitate children having their own voice in the decisions that are made about their lives (Seballos et al., 2011). Treading sensitively is crucial in these situations because people with different cultural expectations could be offended if they felt practitioners were denying their parental rights to raise children as their traditions dictate (World Vision, 2009).

Social workers could assist in smoothing local relations in the Central Asia region by undertaking cross-cultural work and engaging with both majority communities and those from minority religious and ethnic communities. Their contributions could be very useful in the highly disputed Ferghana Valley that straddles Tajikistan and Kyrgyzstan. Social workers have been involved in training immigration officials in human rights to enable them to lessen ethnic tensions by treating (im)migrants with respect and dignity when they cross borders.

Other complications in this region relate to the presence of other players, besides these five key ones, interested in water allocation and use. Afghanistan, Iran, the Russian Federation and China all have strategic interests in the distribution of water in Central Asia. The views of the latter two become particularly significant as they have funds and resources to invest in developing water resources, including for their own use. They also have plans to divert major rivers and potentially cause further substantial environmental damage. Of special concern in this regard are Russia's Siberian Diversion Scheme, and China's diversion plans for the Ili and Irtysh rivers which it requires for the development of Xianjiang Province and resettlement of a further 40 million people in this territory (Allouche, 2007). Were all these plans to come to fruition, this could lead to the complete destruction of the Aral Sea. Were their additional demands on the finite resources that exist here to result in disputes involving all these players, enormous additional suffering among peoples and further extensive damage to the environment would result.

The water disputes in Central Asia illustrate how failing to agree measures for sharing resources and protecting the environment by reaching accords that all players will respect can cause endless havoc. Sharing scarce resources equitably and agreeing formulae and procedures for implementation, and mechanisms for monitoring their observance, go hand in hand with peace.

Case study

Social workers have been involved in solving water disputes in other ways. Some participate in civil society organizations or NGOs working on these issues in Central Asia. For example, the Fund for Tolerance International (FTI) is a Krygyz NGO that was set up in 1998 to facilitate the peaceful resolution of disputes and create links between countries in Central Asia, especially those sharing borders that touch on the Ferghana Valley (Tajikistan, Uzbekistan and Kyrgyzstan), where disputes about borders and water have erupted into violence a number of times. FTI has well-developed links to Kyrgyzstan, and attempts to monitor borders and the people crossing them to ensure that those entitled to cross can do so safely. Its various activities have included working with young people and staffing the camps for internally displaced people after the 'Small Batken War' in southern Kyrgyzstan in 1999 (Allouche, 2007). It has, therefore, had small successes that could be expanded further. The FTI is seeking to do this.

Social workers involved in conflict situations have to consider whether or not they have a vested interest in the outcomes. This includes whether they belong to one of the ethnic groups involved in the dispute; and whether they are prepared for the stresses that this type of work induces in the worker, especially if they live in the disputed area or near it. Ramon et al. (2006) suggest that social workers caught up in such situations can do their work better if there is a legislative framework from which to operate, such as the Good Friday Agreement which pertained in Northern Ireland, one of the locations they describe in their article. Northern Ireland had several other relevant features to be taken into account. These were a history of sectarian violence between religious groups; the existence of parallel systems of social services, often involving parties to the dispute catering for their own supporters alongside state-provided services; and the backdrop of colonialism that had transcended time, but was firmly rooted in place and identities. Besides drawing on the processes of reconciliation and being aware of these elements, social workers need to draw upon their traditional skills of listening to others, reflecting on information that they have collected, affirming people's concerns, identity and sense of place, and building on their commitment to the environment, which has usually been severely degraded by violence perpetrated by all parties to a dispute. In Northern Ireland, Mariead Corrigan Maguire and Betty Williams, with the help of then-journalist Ciaran McKeown, began the Women for Peace movement, renamed Community of Peace People, and now Peace People. Both Williams and Corrigan Maguire worked with children and received the Nobel Peace Prize in 1976 for their peace work.

Consumption of non-renewable fossil fuels

The consumption of non-renewable fossil fuels has become a greater concern now that these are recognized as finite resources that cannot cope with constantly rising global demand. Demand is outstripping supply, and without the substitution of substantial renewable forms of energy, many analysts are predicting that serious economic and social problems will ensue. Like water, fossil fuels, including oil for petroleum, are not distributed evenly across the globe, and price is used to determine both their value and who can afford to buy them. Trade in these commodities has become the major means for distributing them. However, the market-place cannot distribute scarce resources equitably, and other means for those who do not have the money to access expensive commodities must be found. Nor can a market that prioritizes profit motives make the value judgements that give precedence to non-commercially driven courses of action that could or ought to be followed, e.g., leaving oil reserves in the ground instead of exploiting them.

Such considerations become important in places like Canada where the controversial extraction of oil from the Athabasca bitumen deposits, or tarsands as they are commonly known, is proceeding apace. A 'secret' report released through access-to-information legislation claimed that high levels of pollutants including hydrocarbons and heavy metals were present near mining sites and in the waters of the Athabasca river (De Souza, 2011). Although the Canadian environmental movement has protested strongly against this development, the country's Prime Minister, Steven Harper, insists that their claims of environmental pollution are 'exaggerated' and that they should be grateful for the 100,000 jobs created by an industry that is predicted to contribute CDN$1.7 trillion to the economy over the next twenty-five years (De Souza, 2011: A26). His framing of the narrative neglects the: wide range of jobs that could be created by promoting renewable energy sources; costs of the devastation to the environment, society and ancestral heritages of indigenous peoples in the area; and damage to soil, water flows, air quality and wildlife which extend over a much wider part of Western Canada. Harper also withstood daily criticism of the contributions that this particular extraction of oil has made to climate change, and the 'Fossil of the Day Accolade' awarded to Canada for being the most polluting nation on many occasions at the side events in successive UNFCCC meetings, including COP 17. His reluctance to engage with environmentalists' concerns is consistent with seeing nature's bounty as a commodity for exploitation, because these tarsands constitute one of the largest known oil reserves on earth. Deferring their extraction to some future point when technology might be able to deliver this commodity in a more sustainable manner and at less cost to the peoples, flora, fauna and planet is unthinkable in market-place logic, and so enormous quantities of water and energy are utilized to lay bare the land's largesse for the few people who profit from the endeavour.

Another unsustainable extraction process involves removing natural gas from shale through a procedure known as hydraulic fracturing or 'fracking'. In this process, chemicals that are pumped down into rock to release shale gas could enter the water table and have given rise to concerns about: safeguarding the supply of potable water (Hagerty, 2011); the release of methane, a potent greenhouse gas; and earthquakes (Leggett, 2011). As in the case of Canada's tarsands, the economic stakes are high. Shale gas now accounts for one-third of the supply of natural gas in the US, and has lowered energy costs by bringing down the price of natural gas from US$15 per British thermal unit (BTU) to US$3.20. The use of shale gas as an energy source and in manufacturing processes has lowered electricity prices and promoted growth in the petro-chemical industry, which includes the manufacture of plastics, and the steel sector (Hagerty, 2011: B8). Technical analyses of this activity downplay the damage borne by the physical environment and on people's health. Social workers have an important role to play in opening up discussions so that they take account of the costs carried by all stakeholders, not just those calculating their profits, and enable communities affected by such initiatives to make well-informed decisions about what constitutes sustainable developments for them.

As in other market-based approaches to accessing resources, people with the lowest income have least access to fossil fuels, unless countries subsidize them. Some analysts have argued that too high a state subsidy has been responsible for the failure of some consumers to restrain their consumption of these commodities, regardless of their social standing or wealth because all groups are eligible for state subsidies. Moreover, their studies allege that the middle classes benefit more from such transfers than poor or marginalized people because consumption of these goods dents the budgets of monied groups to a lesser degree than those of poorly paid ones or those lacking an income (Allouche, 2007). Social workers can collect information that supports arguments for more inclusionary approaches to resource distribution and consumption.

Hugo Chavez, as President of Venezuela, targeted one state subsidy on making affordable energy available for poor peoples in his part of the world, among whom indigenous people were over-represented, and thereby shared the wealth produced by the oil-rich country. Included in Chavez's initiative were fuel subsidies to eighteen Central American and Caribbean countries in Petrocaribe (Walter, 2008). Within Venezuala, wealthy elites and opposition political parties that rejected Chavez's stances on socialism opposed this policy. Trinidad and Tobago, an oil producer in its own right, and Barbados also refused to participate in the scheme. Whilst Chavez's project provided oil to some excluded groups, it relied on a high price for crude oil being paid by others. By 2011, the price of oil had dropped substantially from its high in 2008, and endangered the continuation of this policy. If the world price of oil is low, it costs the Venezuelan state more to subsidize consumption. Price has a huge impact on the availability of such commodities because the state's capacity to increase payments to cover the same number of poor people is limited when the wider public cannot pay more. Thus, Chavez may have to reverse his policy to protect Venezuela's financial position in the longer term.

Coal, oil and nuclear energy are subsidized as factors of production by industrialized countries in the West, and of consumption by industrializing ones in the Global South, while subsidies aimed at renewable energy sources are low. Burning fossil fuels discharges 6 billion tonnes of carbon into the atmosphere yearly, but the oceans can only cope with one-third that amount (De Moor and Calamai, 1997: 35). Having and using more energy-efficient homes and appliances would reduce energy consumption and the amount of carbon these discharge. Such technologies must be affordable so that those on low incomes can replace existing machines that have high energy demands and are inefficient with cleaner equipment using less energy. Fossil fuels used for domestic and industrial transport account for 20 per cent of carbon emissions in OECD (Organisation for Economic Co-operation and Development) countries. Significant contributions are made by road vehicles, including passenger cars and road freight, and planes. Cars and planes are high on the consumption scale, with passenger mileage increasing for both forms as people become more mobile. These contribute to climate change and pollution, exacerbate poor health and increase demands for land to build transportation links and parking facilities. At the same time, poor transportation links can distort development opportunities and delay economic advances. For example, poor transportation infrastructures in many African countries have impacted badly on economic development. And the continent remains the world's least industrialized region (Hoogvelt, 2007).

Deforestation and forest degradation have their own energy and pollution costs. Compensating developing countries for forgoing the development of rainforests and other virgin lands is not covered by the Kyoto Protocol, and difficult under REDDs (Reduced Emissions from Deforestation and Forest Degradation). Deforestation, particularly from countries in the tropics, contributes 20 per cent of global carbon dioxide emissions (Pisupati, 2004). But, as part of the Bali Action Plan, incentives for not engaging in deforestation were examined and REDD Plus was agreed. REDD Plus proposes compensation for sacrificing current usage of forest products, and so is a form of payment for not exploiting resources for commercial purposes and reducing greenhouse gas emissions by this means. The actual merit of REDD programmes remains disputed (Angelsen, 2008; Peskett et al., 2008).

Similar issues of protecting rainforests arise in temperate regions like the west coast of Canada and north-western United States, where temperate rainforests abound. Disputes over their protection have been ongoing for several decades, with many environmental groups rallying to their defence. These actions have involved social workers in various roles including as activists themselves, professionals dealing with the consequences for those arrested, evicted or hurt in the conflict, and citizens lobbying elected representatives. For example, in 1993 in Clayoquot Sound on Vancouver Island, Canada, people climbed trees to prevent clear-cut logging of ancient temperate rainforest on lands which has trees that are extremely old. Previous logging had already caused much devastation to salmon streams and habitats in the area, and eco-aware individuals, organizations and groups sought to put an end to such practices. Some activists in the Peace Camp that they created as the base for their protest were arrested and charged with trespass. This demonstrates that protecting the environment carries costs that some people might not find acceptable. This action also exposed how companies who claim to provide jobs for an area pit one group of people against another by creating a binary of those who want jobs and those who want to protect the environment at the expense of jobs. Yet, forest resources are ultimately irreplaceable, given the number of years, or centuries, that it takes trees to grow to the size of those at the centre of this dispute.

The caring triangle in conflict situations

Situations of conflict over scarce resources that are claimed by more than one party raise the issues of others caring about and caring for social workers, alongside their own responsibility for self-care. Caring about and for employees involves employers' responsibility of care towards their workers, even though the employer–employee relationship can be complicated by macro-contexts that neither has the capacity to influence. In the case study below, long-standing conflicts over land claims are intricately woven into everyday life practice for both practitioners and service users.

Case study

Claudia was a newly qualified Canadian social worker of Jewish origins. After graduation, she thought she would live in Israel for a few years to work with Palestinian Arabs, because she heard there were insufficient workers in the country willing to provide them with services in the context of the *Intifada* (Palestinian resistance to their lands being occupied by Israelis). Claudia felt she would be able to empathize with Palestinian Arabs because her family had escaped to Canada following the Pogroms in Central Europe. She had heard her grandparents' stories about their survival so many times that she felt she had lived their suffering herself and could connect with anyone feeling oppressed.

Claudia looked forward to her new post and turned up full of excitement on the first day. She was shocked to find a shabby office on a state of high alert because a suicide bomb had gone off in Jerusalem a few days before her arrival. Claudia decided that she had much to learn and began examining the legislation, office procedures and resources at her disposal. Her first service user was a seventy-year-old man who had lost four sons in the *Intifada*. He started to cry when Claudia asked him about them. He cried even more when he told her his wife had died of grief over this loss. He had one daughter, but she was in England somewhere and rarely got in touch. He needed accommodation because his brother-in-law's wife had had another baby, leaving no room for him in their house. Already, there were five people sleeping in the one bedroom. Claudia had no housing for an older Palestinian Arab man on the lists that she could access. When she enquired of her colleagues, they confirmed there was none. Claudia spent the next couple of hours trying to find somewhere for him, using her ingenuity, but without success. The rest of the day proceeded in a blur of similar stories.

She ended the day feeling very despondent. She did not know what she should do and wished she had someone to talk to. She did not feel she could confide in her new colleagues because they seemed to have acquired a 'hard skin' that protected them from feeling useless in the circumstances they faced. Claudia also felt that she was being made to take sides, instead of being able to act as a neutral professional. This left her feeling uncomfortable, not least because she felt that the old Palestinian Arab man had not been treated fairly in the conflict between Palestinians and Israelis, even though he had lived in

Israel since the Israeli state had been formed in 1948. He had even showed her that he had retained the deeds to his former home. This was a moth-eaten document that showed he owned the land that currently housed a Jewish settler family. She wondered how his citizenship rights as a Palestinian Israeli were being met. However, she was a realist and knew that these issues were too big for her to deal with. While her training indicated that macro-level issues had to be dealt with by politicians who carried the mandate of the people they represented, she was worried about feeling unable to function at the micro-level of intervening with individuals and families, because the context in which she was working and lack of resources were not conducive to practice within it. Moreover, she felt privileged. She could return to Canada and look for a job there. She had so much more than the old man she had seen. He had never had an option.

Claudia's case raises profound issues for social workers who support people in conflict situations to consider and prepare themselves for. The likelihood of feeling affected by the politics of disaster, and being disempowered by them, is high. As Ramon et al. (2006) argue, social workers are unlikely to challenge the causes of violence between different ethnic groupings. They seek comfort in professionalism, especially its commitment to the technologies of the tasks to be undertaken, rather than address the complex moral and ethical questions posed by being in such situations. However, this choice was not open to Claudia because her own social work values were contradicted by the realities within which she had to practise. The political context was particularly unhelpful. By not resourcing the office sufficiently to enable practitioners to assess needs and respond effectively, the strong message was given that the service users involved were not valued despite their legitimate claims to services. Additionally, this case study raises important issues about the care of social workers caught up in potentially violent disputes over scarce commodities, land in this case (although there are also water shortage issues in this area, which I do not have time to consider). At the micro-level of practice, green social workers have three parties whose needs have to be addressed in such work: the peoples involved in the dispute and who present themselves as service users; the environment (in all its aspects); and themselves, for whom the duty of care rests with both the individual practitioners and their employers.

Social work education is responsible for addressing these issues in the curriculum so that practitioners can be enabled to cope with realities on the ground (Ramon, 2008). Loneliness and isolation among service users and practitioners are important areas to be covered. Besides learning about the historical legacies in such situations, social workers should be prepared for encountering many problems that they would normally face in the course of their work, e.g., disability, mental ill-health, youth justice problems; and foreseeing that these may also loom large. They should acquire generic skills and specialist ones linked to working in conflict situations where violence is normalized as the backdrop within which practice occurs.

There are large numbers of Palestinian refugees in neighbouring countries in the Middle East. As time passes, it seems unlikely that Palestinians will be able to return to their previous lands. Some 320,000 Jewish settlers have now made these their own, even though their settlements have been built in contravention of international law (Ramon et al., 2006). These realities are likely to increase the stress that practitioners working with either side will experience. Additionally, they have to be prepared for being caught up in state surveillance and disrupted transportation, communication and energy systems at any point during the day or night. Developing one's own personal coping strategies is a worthwhile pursuit and should be considered before beginning to work in any locality marked by conflict. For some practitioners, this will include finding ways of distancing oneself from the work. For others, it will be 'othering' those who are different so that empathy is reserved for those who are like oneself. An ever-present, high-quality mobile phone becomes indispensable in keeping track of people's whereabouts, and can provide a degree of certainty in uncertain circumstances. Collective or team support is also essential. Forming support networks can be liberating, especially if these can be accessed readily when needed, to discuss various concerns that arise in practice and explore one's feelings about these.

Responding to expressed human needs in conflict situations raises difficult issues. From a green social work perspective, these reveal a caring triangle involving residents, scarce resources (land and housing) and practitioners who become embedded in micro-practice. Unarticulated emotional needs that hover in the background make routine practice even more complicated. Claudia's position and responses also demonstrate the importance of having a context that is conducive to working with individuals and families. A holistic response would require a series of actions, not all of which could be undertaken by an individual practitioner but should be attended to by the social work team as a whole. These include forming alliances to promote transformative social change with a range of stakeholders outside of the profession such as politicians, the general public and media.

A wider alliance would enable social workers to raise issues about the macro-context which requires a political solution. Social workers could be involved in lobbying for the peaceful resolution of this conflict. At the level of micro-practice, the social work team would argue for resources that could meet the needs of service users. In the case above, housing is a crucial one. The team would have to respond to the other needs that are being presented, including the emotional grief being experienced by the old man. This raises questions about the appropriateness of the model of an individual practitioner working alone with an individual service user in conflict situations. Finally, the emotionality of practice and its impact on the individual practitioner requires consideration. As Cronin et al. (2007) indicate, addressing the mental health needs of practitioners involved in disaster relief work is a crucial, if neglected, part of practice. Claudia's experience exposes its relevance in her situation.

Conclusions

Climate or environmental justice is embedded in human rights and the dignity of the person. These include the right to decent housing, and compensation if evicted; the right to security of the person and possessions; and freedom of movement, especially within the nation-state (Brown, 2008). Governments have to plan effectively and find funds that will enable them to fulfiltheir obligations under international law and the covenants that they have signed if they wish to avoid social disorder and intensified conflicts, especially as resources become scarce and individual self-help attempts fail to resolve issues.

Including environmental considerations in discussions of poverty enables policy-makers to move beyond thinking purely in terms of the dollars per day required to purchase the basic requirements of food, clothing and shelter, and become more holistic in their thinking about what aspects of social organization need to be changed. Becoming holistic in their approach requires them to cover: access to markets; employment opportunities; empowerment and participation in state institutions; political decision-making processes and governance structures at local level; security in terms of protection against ill-health, food scarcity and water shortages; reducing vulnerability to disaster and economic shock; predictable variations in resource availability; human capabilities; and capacity-building in education and skills (Pisupati, 2004).

Displacement or internal resettlement of populations may occur for numerous reasons, e.g., deliberately planned movements caused by constructing dams, as occurred in England and China; loss of housing due to calamitous events like flooding or earthquakes; and evictions to meet the needs of well-heeled individuals acquiring land in both urban and rural areas. The types of displacement that occur may be temporary or permanent; and can involve the local, national, regional or international domains. Investment in disaster-proofing housing may reduce the number of migrants caused by certain types of disasters, e.g., earthquakes and flooding. However, the record for assisting climate change migrants is not promising, given experiences in Vanuatu, Papua New Guinea, India and the USA (Leckie, 2009).

Social workers can help local people achieve their aspirations in pursuit of a holistic sustainable development that safeguards the interests of people, flora, fauna and physical environment. They can advocate for better policies in this regard, and hold politicians and firms accountable for what they do. Focusing on peaceful ways of resolving disputes over scarce resources and devising ways of sharing these equitably would assist in the creation of a holistic sustainable development.

Co-producing knowledge and solutions to the problems being addressed will facilitate sustainability. Community participation is crucial to successful attempts at developing holistic sustainable environments. This involves community development workers helping people to identify potential strategies that will enable them to: meet their daily life requirements; protect their environment; develop local leadership; observe local customs; and relieve the pressure of their demands on local environmental resources, e.g., drilling local wells that can provide water to irrigate crops and planting those that are less water-hungry than others. Community development workers can assist local communities to conduct their own research into water consumption and wastage and energy usage to provide data for arguments that seek to change existing arrangements.

9

Interrogating Worldviews: From Unsustainable to Sustainable Ways of Reframing Peoples' Relationships to Living Environments

Introduction

Western ways of thinking about the world and industrialization processes have focused on hierarchical and binary views of life that place people and their interests at the top of its hierarchy. This binary pits 'man' against 'nature', with 'man' aiming to control and exploit nature. This approach has been typified as modernity. Moreover, this worldview is deemed superior to others and glories in being the product of rational thought processes having a firm foundation in empirical evidence and scientific discoveries. Whilst the West has benefited from such depictions of industrial reality because it has raised people out of the forms of economic deprivation that marked the nineteenth century, the model has failed to eradicate disadvantage and marginalization within its own borders, and intensified poverty elsewhere by destroying non-industrial lifestyles and promoting underdevelopment in the interests of commandeering the Earth's natural resources for its own projects. Wealthy elites in other countries have similarly benefited from the proceeds of industrialization and the adoption of Western lifestyles.

Alternative worldviews have existed and continue to do so. Significant among these have been indigenous approaches to life which can be found among the aboriginal peoples already in the West and those elsewhere who have struggled to keep alive their cultures, languages and traditions despite the onslaught upon them legitimated through colonization. Other indigenous peoples in Asia, Latin America and Africa have sustained traditional lifestyles, against the odds. Like their counterparts in the West, they are demanding the restoration of their rights over their resources and more sustainable lifestyles based on a symbiotic and respectful relationship with nature. Those involved often engage in indigenous movements or collective struggles which have had a significant impact upon social work. Key among their approaches to daily living is the integrated relationship that they envisage among themselves as people, and between them and their social and physical environments. Their conceptualizations of themselves as keepers of the flora, fauna and Earth's natural resources for current and future generations are embedded in a spiritual connection between people, other forms of life, and inanimate objects.

Some people refer to this as a spiritual orientation to living things and their physical environments. These approaches are also typified by caring relationships that link individuals, (extended) families and communities in a mutual responsibility to care for each other. These connections have enabled such groups to survive attacks on their ways of life and physical environments. The question of whether such connectivity can be maintained and sustained within Western modes of industrialization remains for further research. Welfare states in the West theoretically embodied this altruistic potential. Moreover, in an age of austerity, extensive reductions in public expenditures threaten to undermine the capacity of informal caring relations, expressed as care services provided through both the voluntary sector and domestic economy of the household, to fill the gap left by the loss of publicly funded facilities available to all, free at the point of need. This turn of events is likely to intensify the caring burden borne by women, who are primarily responsible for caring, whether paid or unpaid.

In this chapter, I consider how indigenous beliefs, particularly those of the First Nations in Canada and the Maori in Aotearoa/New Zealand, have reframed social work practice as they sought to overcome colonialism, racism and genocide, and draw out lessons for practitioners. Indigenous worldviews tend to be collective and aimed at causing the least possible disruption to the natural environment while leading sustainable lifestyles that are based in local communities. Their knowledge can yield insights that might help social workers working in densely populated urban areas to: assist city dwellers in reconnecting with the physical world; enable people inhabiting rural settings to promote endeavours that modernize or industrialize agriculture in sustainable ways; and prevent the mass migration of young people from rural villages to the urban centres of large cities to earn their livelihoods.

By going into cities, often without members of their family or band/ tribe to support them, indigenous peoples encounter contemporary forms of discrimination and oppression. Dominant societies have yet to address diversity effectively in either urban or rural areas, and the plight of indigenous people in cities and the countryside is testimony to this social failure. The absence of culturally appropriate facilities and support networks to assist those migrants who arrive in cities and rural areas to establish worthwhile and sustainable livelihoods is an issue that social workers could address. They could thereby mediate the transition of migrants into new life spaces. In intervening in this matter, social workers would have to explore the macro-, meso- and micro-levels of practice. The macro-level would include the specific history of colonization as it applied to these particular groups. The meso-level would include policies and legislation that would enable practice to be carried out with this group and facilitate the acquisition of resources needed to provide the necessary services. The micro-level would involve responding to the needs of specific individuals and families, and would require detailed knowledge of the specific band and culture of that individual.

At this juncture, if non-indigenous social workers were not skilled in working with indigenous people, an indigenous social worker from that community might have to be brought in to work with the person. There is also an obligation on non-indigenous practitioners to work with indigenous social workers and their organizations to make such interventions culturally relevant and appropriate to the needs of those wanting to use the services provided. Knowledge of the social relationships practised within indigenous communities and different resources embedded in their traditions is useful too. A partnership that can address issues of diversity within a framework of equality (Panet-Raymond, 1991) requires the social work team to establish power-sharing relationships between practitioners and service users, and address practice at all these levels. In the process, social workers from the dominant social group would have much to learn about alternative approaches to life, if they are to ensure that their practice is culturally aware and appropriate for use with particular diversities.

Indigenous Worldviews

There are an estimated 370 million indigenous people on Planet Earth. The term 'indigenous' is a contested one and there is no agreement as to its meaning. I have written about this issue earlier (Dominelli, 2000) to highlight the association of colonial power relations and notions of inferiority with the words 'indigenous' and 'aboriginal'. As there are no useful alternatives, I have retained these terms in this book. Essentially, I use these interchangeably to mean people who have retained ancient cultures and traditions despite the onslaught upon them of colonization and modernity expressed as capitalist industrialization in its various guises.

Like other colonized groups, indigenous and aboriginal peoples have had their own unique and distressing experiences of discrimination and marginalization as a result of colonization linked to capitalist forms of industrialization involving various European powers, including France, Spain, Portugal and the UK. I cannot cover all their stories here, but I select those that have impacted significantly upon social work practice in the English-speaking world – the First Nations of Canada and the Maori peoples of Aotearoa/New Zealand. I also refer to one indigenous group in a Spanish-speaking country and the international conventions that can support social workers who wish to enhance their practice in working with indigenous peoples, learn from their insights and incorporate these into holistic worldviews that encompass relationships between people, plants and animals and the physical environment as an integral part of their social organization of society. The most significant of these was the Declaration on the Rights of Indigenous Peoples, which was adopted by the UN General Assembly in 2007. The political alignment in the vote was interesting for what it revealed about various interpretations of the provisions around the recognition of indigenous peoples as having full and equal citizenship rights wherever they lived. Those with active indigenous movements, namely Australia, Canada, Aotearoa/New Zealand and the United States voted against the Declaration in General Assembly deliberations. The Russian Federation, which has its own indigenous peoples to empower, abstained. The Nordic countries that have within their borders a significant proportion of the indigenous peoples of Europe, primarily the Sami, voted in favour.

Work to secure the passage of this Declaration began in 1982. The Working Group of Indigenous Peoples (WGIP) created under the auspices of ECOSOC (the UN Economic and Social Council) worked hard to reach this point. The WGIP passed on its deliberations to the UN's Commission on Human Rights in 1994. It proved to be extremely contentious because its provisions included the affirmation of indigenous peoples' claims not only to their cultural traditions and languages, but also to their traditional rights to land and its mineral and other resources. It went through a number of versions, but was not agreed by a majority of those sitting on the Commission until 2006. Although passed by the General Assembly, this Declaration is more a sign of intention than a legally binding instrument, so there is much further work needed to ensure that countries ratify a Convention that can then be incorporated into national laws. Conventions are more authoritative and have greater potential in assuring indigenous peoples that their status will not be minimized or result in their being treated as holding an inferior status to those who already have Conventions addressing their needs, e.g., women and children. Although women and children of indigenous descent are also embraced by these other conventions, their coverage does not extend to all aspects of an indigenous identity, including their heritage rights and desire to retain practices reflected in traditional forms of hunting and fishing.

First Nations peoples in Canada, Turtle Island

Canada's indigenous or aboriginal people include the First Nations, Inuit (formerly called Eskimos) and Métis (those with a mix of (primarily French) European, and First Nations origins) peoples. Contemporary Canadians honour them through National Aboriginal Day. It was proclaimed in 1996 as 21 June, but not all provinces have accorded this date the standing of a statutory holiday. This lukewarm response symbolizes the unfinished business of reconciling the interests of Canadians of aboriginal heritages with those of European ones, and increasingly other ethnic groupings. Some First Nations peoples refer to Canada as their territory in 'Turtle Island' (North America). The term 'First Nations' is used to depict the historical arrival of this group as the first residents to inhabit this land, and has replaced the European misnomer of 'Indian'. They are made up of bands or tribes, which form the basis of their identity, e.g., Coast Salish, Anishinabe. Each Nation may be small in numbers because many of them were wiped out through diseases, armed conflict and deliberate strategies on the part of Europeans committed to eradicating their values, languages, traditions and ways of being (Haig-Brown, 1988). There are 1.2 million people belonging to 600 recognized bands currently comprising the First Nations in Canada, with approximately half of them living in British Columbia and Ontario. Each has its own language, culture and artistic traditions. Poverty and marginalization have featured strongly in their history. The Inuit, whose homes have traditionally been in Canada's Arctic North have recently formed the autonomous region or territory of Nunavut in which they form the majority of a small population, whilst remaining within the federal juris-diction of Canada. There are many outstanding disputes over the treat-ment of indigenous peoples, their self-determination and sovereignty, land, mineral and water rights across the Canadian federation.

First Nations peoples speak from their location, as people of a par-ticular territory or geographic location, and gain their insights from their relationship as custodians of the piece of Earth that pertains to them. Their knowledge has been expressed through the 'Medicine Wheel' (Cyr, 2007; Green and Thomas, 2007) and ancient narratives such as the story of creation and the eight fires (Bruyere, 2010). Using these narratives as tools, First Nations scholars have formulated their own theoretical per-spectives. For example, Gale Cyr (2007) has developed an Indigenist, Anti-Colonialist Framework for social work practice. This approach deconstructs capitalist social relations and explores their impact upon traditional First Nations' worldviews. Conflict mediation is a crucial element within this indigenist framework. Although First Nations are fairly specific about their relationship with those of European origins, their peoples have full agency. That is, they can choose to follow the teachings of their Elders, or they can ignore these. Those with European origins alongside their indigenous ones often reject indigenous ways of life as a strategy for surviving in a racist and colonizing society. Negative depictions of their lives are often internalized by First Nations people. For example, Daryl (fictitious name), in a research project that involved fathers of First Nations origins said:

> I was so ashamed of my family . . . of what I was . . . I didn't want to be an aboriginal. My best friends were not aboriginal . . . Everybody thought we weren't aboriginal. So I pretended I wasn't till I messed up one time. (Dominelli et al., 2011)

During his youth, Daryl sought to be accepted by white society and 'fit in' with his peer group, which, with the exception of him, was made up exclusively of young people of white European origins. His cover as a 'white' person was blown when he accidentally let slip that he under-stood what a group of First Nations men speaking their own language were saying.

Daryl had good reasons to be concerned about identifying himself as an aboriginal because First Nations children are over-represented in the child care system, and their young people and adults are over-represented in the criminal justice system. They are discriminated against in the employment and housing stakes in the wider society. And they struggle to overcome the impact of colonization and having been abused through the residential school system wherein they were denied the use of their language, links with their families during school terms, and traditional religious rituals and observances (Haig-Brown, 1988; Grande, 2004). This treatment has caused untold suffering among First Nations peoples and is reflected in high levels of alcohol and drug misuse. One reaction to this position has been to turn the addiction response into a major arena in which indigenous social workers assist in the healing process. Another issue in which indigenous social workers intervene is with children. First Nations have sought to reclaim their more holistic and communal ways of working with children, and have succeeded in forming 28 First Nations agencies in British Columbia to provide services for children and families 'off-reserve', or outside the particular reservations to which First Nations peoples have been consigned by the Indian Act of 1876 (Green and Thomas, 2007). These services focus on providing children with the support of their communities and Elders, and in developing self-confidence over their identities, traditions and cultures.

The Indian Act infantalized First Nations peoples and made them dependents of the federal government of Canada. Many of their traditional rights were lost through this legislation. For women, this was particularly dire, because they lost their citizenship status as Indians if they married a non-First Nations person or lived off-reserve. Moreover, the reservation to which they were assigned may not have been in their traditional territory, and so the very important links that First Nations had had between place and identity had been broken. Thus, First Nations people are dealing with the devastation caused by the (hu)man-made disaster called colonization. To reclaim their culture, heritage and language in a culturally appropriate manner, they have drawn upon their Elders' knowledge, which has been encapsulated as the 'Medicine Wheel' (Green and Thomas, 2007).

The 'Medicine Wheel' encompasses all life and the relationships between indigenous peoples and their environments. Working in a clockwise direction from the apex, it is organized with the North at the top and then moves to the East, then the South and ends up in the West. Each of these four directions contains: the stages of the life cycle; the four different components of human life, namely, the spiritual, emotional, physical, and intellectual dimensions; the four seasons; and the symbolic representation of all 'races' through the use of the colours white, red, yellow and black. First Nations are taught that the 'East', the spiritual realm that is linked with the colour 'red' and the sunrise, represents spring and infancy. This is where they begin to examine and analyse the history of indigenous peoples in Canada, including that of colonization (Green and Thomas, 2007). If work using the 'Medicine

Wheel' is explored clockwise from the top, the North reflects the intellectual realm associated with the colour white, the wisdom of the Elders and old age. It is a time of winter and reflection. It is also a time for dreaming and envisaging a new and hopeful future of fulfilment. Opposite the North, is the South or the emotional realm. It symbolizes summer, youth, the colour yellow and highlights the search for best practices. The West is where the physical world is located. It is associated with autumn, the colour black and adulthood. It is also the location of the 'self', where the work of strengthening families and sharing knowledge occurs.

Children in First Nations families are deemed 'precious gifts'. They form a crucial link between generations, bringing together past, present and future, and are an inherent part of the community which is based on their tribe. This community is also responsible for raising and rearing their children. The focus is on developing children's strengths and talents, with the adults, especially the Elders, acting as role models. Children need this help to grow strong and overcome the racism meted out to them because they live in Canadian society. Their parents often need assistance because they have faced decades of trauma which have undermined their confidence in themselves and their parenting skills (Green and Thomas, 2007).

First Nations' social welfare provisions

Welfare provisions in Canada are subject to provincial jurisdiction and so vary from province to province. First Nations peoples are also subject to the provisions of the Indian Act, which is federal legislation and applies to all First Nations people across the country. The Indian Act was considered both racist and sexist in its depiction of First Nations peoples (Churchill, 1998). This legislation configured them as dependents of their white rulers and subject to state control. It also excluded women from passing their indigenous ancestry to their children if they left their reserves or married a person without First Nations origins. Despite these barriers, First Nations peoples have challenged their depictions as incompetent citizens through numerous struggles (Wiebe and Johnson, 1998). This included challenging the Indian Act by taking the issue of its discrimination against indigenous women to the UN and demanding the restoration of their equal status, which they successfully obtained.

In many parts of the country, there has been a revival of First Nations' traditions and teachings retained by their Elders (Maracle, 1993). These developments have validated indigenous knowledge and ways of learning, and restored self-confidence and autonomous forms of action. First

Nations developed their own theoretical concepts and understandings and applied these to indigenous welfare systems, including those for formulating their child protection systems and assuming control over their operation. They have also become more politically engaged, including playing major roles in national initiatives. For example, First Nations negotiators in the Meech Lake Accords refused to sanction this legislation because it disregarded indigenous peoples' concerns about their rights and heritage and failed to respect their views and positions. They are credited with its demise.

First Nations people have been marginalized from decision-making in the wider Canadian society from the beginning of Canada's colonial history. They have held few parliamentary posts, although this has begun to change recently. For example, Elijah Harper, a member of the Cree Nation, was the first 'Treaty Indian' to be elected to the Legislative Assembly of Manitoba as Member of the Legislative Assembly (MLA) of the province. In 1990, he refused to agree to the Meech Lake Accord aimed at settling Quebec's grievances in the revisions to the constitutional arrangements of 1982 when Canada repatriated its constitution from Britain. Although the federal government had reached agreement with other provinces in 1987, Harper rejected it because it had ignored the interests of First Nations peoples and excluded them from the discussions. Elijah Harper also opposed the Charlottetown Accord which followed, even though Ovide Mercredi, Chief of the First Nations Assembly, supported it.

Case study

Elsie was a Canadian social worker of white European heritage. She had made good friends with a number of First Nations people living in her community and learnt to appreciate their traditional ways of living and to love their culture. Over time, she changed her profession from being a school teacher to a social worker. She liked her colleagues and found that the team's work reached high standards in all areas, except that no one worked with service users of First Nations' origins, on the grounds that they were not 'competent to do so'. She sympathized with this stance because she realized she worked in a team that did not employ any First Nations or other indigenous social workers. They had neither been trained in such work, nor had any experience of working with this clientele. Because the team knew of Elsie's interest in First Nations cultures, her manager always gave her the cases that involved First Nations people whether children or adults. Elsie went along with this for some time, and she always checked what she was doing with the Elders in the community in which she resided. At one point, one of the Elders asked her, 'Elsie, what are you doing? How far have your colleagues learnt how to work sensitively with First Nations individuals, families and communities? How much have they learnt about our cultures?'

Elsie blushed because she realized, the moment she was asked these questions, that she had colluded with her colleagues' failure: to take responsibility for living in a society that included First Nations peoples; to learn to respect them as individuals and communities who shared the same land together; to value their contributions to society; and to gain insights from their enormous font of knowledge and their wisdom. By working with all cases involving First Nations peoples, she had ensured that the team took seriously neither the employment of nor the imperative to learn about First Nations peoples. Although she had many discussions with individuals, trying to persuade them to join her – none had.

Elsie promised the Elders that she would raise this issue with her manager and ask for a team meeting in which they could develop a strategy for training all the social workers in indigenous cultures and history, so that the existing state of affairs did not continue. The Elders agreed to support her in this task to ensure that none of the children or adults of First Nations origins suffered from receiving inappropriate services in the process.

Elsie's predicament demonstrates the complexities of becoming involved in working across cultural divides. It also reveals how easy it is for practitioners to fall into the trap of feeling that they have to do everything themselves in order to ensure that those entitled to receive culturally relevant services can do so without having their well-being put at risk. This solution is not sustainable. Workforce diversity remains elusive and it is not conducive to the development of other members of the team, because they do not acquire the skills necessary for doing such work. It also indicates the importance of working jointly with First Nations peoples to ensure that the services that are provided in mainstream settings do not cause further damage to indigenous service users. This is very important in a context where cross-cultural work has meant that indigenous men, women and children have paid the price for the poor practices perpetrated upon them. Elsie was fortunate that the Elders were willing to support her in initiating transformative change in her own agency's ways of working with cultural diversity and building on the connectivities that can exist between people who respect and value each other. They could have refused to do so and insisted that only First Nations practitioners work with First Nations peoples. In this sense, they might have been hoping that, with the safeguards that they would institute, no harm would be done to First Nations peoples, and that maybe, if transformative change were possible, it might point the way to lighting the Eighth Fire when all peoples would respect each other's diversity and work as equal partners for the good of all, as suggested by Bruyere (2010).

The Maori worldview

The Maori in Aotearoa/New Zealand also position themselves as the first inhabitants of that territory. The Treaty of Waitangi of 1849 recognized the Maori language and their rights in the country on an equal basis with the Europeans (Pakeha). Realizing these has been difficult, despite Maori enforcement through a history of resistance to being colonized and treated in a racist manner. Their commitment to maintaining their own traditions and cultures has enabled them to withstand the worst excesses of colonialism which resulted in disproportionately more of their young people being taken into care and their young men being imprisoned (Ruwhiu, 1998). The development of the Family Group Conference was their response to this state of affairs and indicated their attempt to assert control over their own social services because the existing provisions had played such negative roles in their lives (Ruwhiu, 1998). The *Puao-te-Ata-tu* (Daybreak Report) of 1986 reaffirmed the Maori peoples as a sovereign and self-determining group who were capable of devising their own practices in the rearing of and caring for children and other family members. The Report was produced by the Ministerial Advisory Committee on Maori Perspectives for the Department of Social Welfare. This Report became a turning point in empowering Maori peoples and enabling them to affirm their culture and language. It also enabled them formally to assert their ways of caring for their peoples. Their determination on this front led to their challenging and reordering the mainstream social services and welfare system of the wider society, and in the case of Family Group Conferences, the wider world (Schmidt et al., 2001). In Aotearoa/New Zealand, developments from these indigenous models were enshrined in the 1989 Children, Young Persons and Their Families Act. Their language was an important part of the struggle to retain their identity, and some Maori words that are central to their worldview have entered the lexicon of both Maori and Pakeha peoples. According to Tait-Rolleston and Pehi-Barlow (2001), crucial ones are:

- kia ora (greetings)
- whakapapa (genealogy)
- whanaungatanga (kith and kinship ties)
- maatua (parents)
- kuia/koroua (grandparents)
- iwi (tribe)
- tupuna (ancestors)
- waka (canoe)
- te ao Wairua (the spiritual realm)

- whenua (land)
- whanau (family)
- tamariki (children)
- mokopuna (grandchildren)
- hapu (sub-tribe)
- Kaumatua (Elders)
- te ao Maori (the Maori world)
- koha kii (gift of words)
- Ngai Tuhoe, Te Arawa, Te Atihauni-a-Parangi (names of tribes)
- Pakeha (people of European descent or colonizers)
- Puao-te-Ata-tu (Daybreak Report).

The Maori worldview is rooted in a respect for nature, and their kinship affiliationsh. Reclaiming their identity has involved long struggles against colonization, and in this respect they have a similar interest to indigenous peoples in other countries. According to Tait-Rolleston and Pehi-Barlow (2001: 229), the Maori vision of society is rooted in 'the interconnectedness of the individual, the family kinship systems, the physical environment and the spiritual realm (te ao Wairua)'. The connectedness between the different parts of their conception of society links the 'intellectual, emotional, physical and spiritual domains' and provides a holistic basis in which every man, woman and child is embedded. Their relationship to and association with the physical environment is what assures Maoris of their well-being (Tait-Rolleston and Pehi-Barlow, 2001). There is, however, some variation between Maori peoples that depends on the ancestral tribe to which they belong. Both their worldview and the specifics of their tribal traditions have been passed down through the generations, mainly by word of mouth. Each person, therefore, has a relationship with Earth and others that transcends time and space. This worldview positions them well in building on the past in terms of valuing and caring for the environment for current and future generations. Children are an important part of this link and valued for this connection and as beings in their own right. These connections, beliefs and traditions form the theoretical underpinnings of Maori social work constructs:

> Concepts of balance, reciprocity, genealogy, the tribe or extended family are central to this [social work] paradigm and directly affects notions of human justice, social justice and the law and what they mean to Maori peoples. (Tait-Rolleston and Pehi-Barlow, 2001: 247)

Moreover, they also influence Maori responses to Western materialist lifestyles and lack of spirituality, which they fear will undermine Maori ideas about how to lead a good life by asserting Western superiority and authenticity above all else, thereby invalidating other peoples' truths if they were different. Tait-Rolleston and Pehi-Barlow (2001) also suggest that a strength of Aotearoa/New Zealand's policy of bi-culturalism is that it may impact upon Pakeha social workers and facilitate the questioning of taken-for-granted assumptions in both their practice and daily lives. This also applies to the Pakeha educational system and practices within it.

Indigenous Peoples' Struggles for Land and Self-Determination

Other indigenous people have also struggled for their cultural heritage and traditional rights. These are too numerous for specific consideration here. But their stories contain consistent threads that link their histories together including colonization, marginalization, destruction of their cultures and languages, and loss of land and other resources. That many have survived this onslaught is remarkable and a tribute to their strength and determination to survive with dignity so that their descendants can inherit their positive worldviews and approaches to the Earth and its beauty. Like indigenous peoples in other continents, those in Latin America had dreadful experiences of colonization and terrible outcomes as a result of the damage inflicted through their encounters with white Europeans. For example, the Taini peoples in the Caribbean were wiped out. Others, like their North American counterparts, came close to extinction as the colonizers fought them to secure vast natural resources encompassing land, gold, silver, copper and other precious commodities. Many were enslaved to work the mines, others on industrial sites and agricultural plantations. When the local population was insufficient for these purposes, the slave trade ensured a plentiful supply of labour from Africa. It was not only the people who suffered from the rapaciousness of colonization. The flora and fauna of these localities also suffered, and many species were driven to extinction (Cox, 2000).

Colonization, by imposing industrial processes and ways of life on pre-industrial societies, plundered the peoples' riches and environmental wealth and destroyed much in its wake. It also upset the balance that indigenous peoples had achieved between people and nature. Some of their insights are evident in today's deep ecology movement. According to Zapf (2009: 3):

> Deep ecology promotes harmony and connection among all forms of being, a mutual dependence rather than human domination of the

natural world for economic gain. Diverse ecosystems have intrinsic value beyond their economic utility for extractable resources.

The Mapuche struggles for land rights in Chile

The importance of having a deep ecological perspective on life is evident in the lives of indigenous people in Latin America. Indigenous peoples' struggles on this continent, along with many others, are long-standing and cover violations of human rights and traditional heritage rights associated with usage of the material bounty provided by Mother Earth in all its diverse elements. These have many insights regarding topics considered in this book. I have selected one that is rarely considered in English-language texts as indicative. It is the example of the Mapuche Indians' struggles to reclaim control over their lands in Chile. Achieving success in such situations is difficult, especially when the state which is supposed to act as a guarantor of indigenous peoples' rights is implicated in their infringement by ignoring its obligations to do so and siding with those seeking the industrialization of wilderness lands, in the name of the greater good. But can the 'greater good' be morally justified if it leads to the destruction of something precious and irretrievable once it is damaged? Surely, care of such resources has to be at the centre of the action. Who decides such issues, in tribal communities where the roles played by tribal elders and chiefs are important, particularly those whereby they are respected as sources of wisdom that must be consulted? The case study below reveals the impoverished outcomes when consultation is tokenistic or fails to engage adequately with the issues and peoples involved. It also highlights how power relations between different ethnic and racial divisions have to be taken on board to achieve egalitarian outcomes that respect peoples and their physical, social, cultural, spiritual, economic and political environments.

Case study

The Mapuche Indians are an indigenous people in the southern Araucania region of Chile. They have resisted successive waves of colonization, from the Incas until the present day. They have been subjected to a strategy of assimilation under the auspices of the state, but implemented mainly through the Catholic church. Assimilation included depriving them of their lands for development purposes over which they had little say. General Pinochet, the dictator ruling Chile from 1973 to 1990, authorized the privatization of Mapuche lands in 1979, adding subsidies and tax breaks as sweeteners to investors. The loss of their lands led the Mapuche to organize collectively, campaign actively for the return of their lands and demand political recognition as a specific cultural group and sovereign electorate. Other indigenous peoples in Latin America were also organizing at this time, and links were made between the different indigenous groups who faced similar struggles over the recognition of indigenous rights and their determination to protect their environment from large-scale development, particularly that taking the form of invasive dams and landfill sites. The indigenous movement was gaining recognition further afield. In 1992, Rigoberta Menchíu, a Mayan activist in Guatemala, was awarded the Nobel Peace Prize for her work with indigenous peoples. The UN named 1993 the Year of the World's Indigenous Peoples, and released the Declaration of Indigenous Rights in that year.

The Mapuche had a particularly difficult experience in getting their voice heard in the context of a long-standing Chilean tradition of a top-down exclusionary form of decision-making called *cupulismo*. The Concertación Coalition government that followed Pinochet promised to include indigenous people in co-participation exercises that involved them in governance structures that mattered to them – but it did not. Consistent failure in the politics of engagement led the Mapuche to form alliances with others sharing their views about the environment and to become more active in articulating their stances, leading to several protests of significant magnitude, especially in response to mega-development projects in their territory involving the exploitation of hydro-power and timber. The proposed development of the Ralco Dam on Pehuenche (a sub-group of the Mapuche) lands aimed to remove considerable acreage from their control and contrasted sharply with the initiatives promoted under Presidents Frei and Allende, whose land reforms returned 70,000 hectares of land to this group of indigenous peoples. Building the Ralco Dam would relocate large numbers of people and cause considerable environmental damage.

The Mapuche were joined in their protests by non-governmental organizations (NGOs) that provided advocacy skills and helped create the Mapuche Cultural Centres to support indigenous peoples who were mobilizing around claims to their culture and traditions. These were formed alongside state initiatives such as the Special Commission for Indigenous Peoples (CEPI). Both organizations aimed to promote indigenous rights. Chile's indigenous peoples were promised much in terms of sustainability and taking care of the environment. They realized little because industrial needs trumped indigenous peoples' rights, although the Environmental Framework Law of 1994 contained many promising features that reflected egalitarian norms including the prevention of environmental degradation, participation, gradualism and the 'polluter pays principle' (Carruthers and Rodriguez, 2009: 746).

The completion of the Ralco Hydroelectric Project, despite Mapuche objections, symbolizes the sense of frustration and failure that the Mapuche experienced in their dealings with various Chilean governments. Under Pinochet, a series of hydro-installations were built in the Upper Bio River Basin. The Pangue Dam was the first to be built; the Ralco was the second. Both were under private ownership and opposed by the Mapuche, on whose lands these power sources were built. Their construction caused major ecological and cultural damage, the industrialization of wild countryside, and forced the resettlement of hundreds of Pehuenche families from the highlands (Carruthers and Rodriguez, 2009). The Mapuche were supported in their opposition to such developments by a range of environmental activists, human rights groups and other civil society organizations, but to no avail. The state was so determined to get these installations completed that it replaced indigenous members of various bodies that would vote against the government line in the ensuing deliberations with those who would side with it. The Ralco Project was completed in 2003.

Similar stories can be told about the private timber plantations on former Mapuche lands. Timber now accounts for 34 per cent of Chile's exports. The struggle to prevent these lands being privatized was fierce, with allegations of police brutality against demonstrators. The Mapuche and their supporters had serious concerns that these plantations would diminish biodiversity, destroy precious habitats, and degrade the environment in general. For example, private eucalyptus and pine plantations require so much water daily that they are drying out surrounding lands and fields. Pesticides applied aerially and herbicides used in agricultural production are contaminating watersheds and undermining people's health (Carruthers and Rodriguez, 2009).

The Mapuche had a range of experts offering advice and support in the long struggles to assert their rights. Some of their supporters died in these conflicts, including Matias Catrileo, a Mapuche student shot by the police. The Mapuche's allies included wide-ranging and diverse groups. They included social and physical scientists, civil society activists, human rights campaigners, anthropologists, social workers, community development workers and engineers. While the Mapuche have commented that not all their supporters worked according to the spirit of co-participation, they have welcomed such contributions to their endeavours to save their heritage and lands (Carruthers and Rodriguez, 2009). The Mapuche's struggles to protect the environment and their cultural heritages remain unfinished business.

The Mapuche case shows how complicated it is to mount successful environmental initiatives when these are not supported by the state. Despite having the support of many people, and being aware of the legislation that they would be able to draw upon, victory is not assured. The power of money to determine what kind of social development is practised at the local level highlights the lack of balance in decision-making between wealthy elites and indigenous people. The state, as the holder of the interests of all of its constituents, does have a responsibility to ask whether it can promote less damaging and devastating forms of development. Would it be possible to produce the energy required to meet the nation's needs in other ways? What role can renewable energy technologies play in this? How can the Mapuche be involved as full col-laborators who work with the state and industry to devise solutions that will be acceptable to all these parties? Is the exclusion of the Mapuche in the deliberations about their lands a condition of racist colonial ideas in which indigenous people were treated as infantile and dependent and therefore unable to make their own decisions? Any government with indigenous people living within its borders has a responsibility to answer these questions and work with indigenous people in good faith to find solutions to very complicated and difficult problems. Innovative and creative responses that actually engage with indigenous peoples' concerns have to be found. Social workers can help by working with people on the ground to devise these new approaches. Chilean social workers have a long history of practice that is embedded in communities and that seeks to bring all parties to a dispute to the discussion table and to move forward together (Ponce de Leon, 2008).

The Mapuche's struggles also highlight the depth to which their mar-ginalization has become institutionalized, even within the environmental movement. They found it difficult to obtain the information they needed to make sound decisions and had to work extra hard to make themselves heard, even in settings where they should have been controlling the dis-cussions (Carruthers and Rodriguez, 2009). This response indicates the importance of environmental activists listening closely to indigenous peoples' views, engaging with them through more participative discus-sion processes, and enabling their own relationships with local residents to become more egalitarian and empowering. Otherwise, they may have one goal like safeguarding the environment in common with indigenous peoples but their practices will replicate the exclusionary and oppressive ones favoured by ruling elites, which indigenous people are more than familiar with and find unacceptable. Environmental struggles have made indigenous people more aware of the issues involved in protecting their environment. Indigenous people understand the complex power dynam-ics that are at play, and know that they have the capacity to undertake and critique environmental assessments that are not conducive to the preservation of their lands. And they know that they share similar struggles with indigenous peoples elsewhere on the planet. Having this knowledge is crucial to their being able to transcend the politics of disas-ter that short-change them. This more comprehensive and politically situated approach to working with environmentalists and other potential allies marks the green social work approach as different from the eco-spiritual perspective advocated by Coates et al. (2006).

Working through the UN and associations that they have established, indigenous peoples have learnt from each other and improved their social capital by extending their networks overseas to gain support from other

indigenous peoples as well as non-indigenous activists, e.g., the Indigenous Peoples' Restoration Network based in Canada; the Indigenous
Peoples' Global Network on Climate Change and Self-Determination,
launched in 2010 at the climate change talks in Bonn, Germany. Additionally, the indigenous peoples of Canada have developed their own
indigenous environmental assessments to ensure that their views are
reflected in discussions about developments in their country, and have
made these available to anyone. Are there insights from these experiences
that would be useful for the Mapuche to know about? I present key
elements of the holistic approaches and processes involved in environmental assessments drawn up and followed by First Nations peoples in
Canada below. The Mapuche can decide for themselves whether these
are useful in their particular circumstances.

A First Nations Environmental Assessment Framework

The First Nations in Canada, like indigenous peoples elsewhere, have
sought to develop tools to help them assert their rights to a safe and
healthy environment that protects their interests as peoples and those of
the ecosystem. Their environmental assessment is a processual instrument that is devised and best used collectively to ensure that all aspects
of a situation are taken into account. First Nations define an environmental assessment as 'a process used to assess and predict the environmental effects of a proposed project or activity before the proposed
activity is carried out' (FNEATWG, undated: 2). The First Nations Environmental Assessment Technical Working Group (FNEATWG) has
devised a toolkit of 343 pages to assist people in the process of assessing
the impact of any proposed development on residents and the physical
environment. The emphasis on process insists that how decisions are
reached is as important as the actual decision itself. Effective participation at the level of the tribe lies at the heart of this tool. The environmental assessment is an integral instrument in creating opportunities
for the sustainable development that will ensure survival now and
promote the resilience that will enable the community to grow and continue to pass on its heritage to its children in the future. The environmental assessment toolkit aims to be holistic, and keeps an open mind
as to what the final decision will be about any particular project. As a
holistic tool, it aims to acquire as much information as possible for
making good decisions. The factors that this tool recommends are taken
into account are:

- *Environmental factors*
These include the quality of the entire ecosystem from the air to the water (fresh water and salt water), and the land in between; from the animals to the vegetation; the location of living and inanimate objects and quantitative measurements about the location of all of these; and the movements of living beings.
- *Health and socio-economic factors*
The demographics of the area, population movements, social structure, community stability, the built infrastructure, quality of life indicators, health risks to the individual and the group, employment, business opportunities offered and forgone, and the adjustment programmes to be introduced are basic items to consider.
- *Cultural factors*
These refer to the specifics of traditional lifestyles, language, customs and resource use (FNEATWG, undated).

The FNEATWG environmental assessment tool has additional purposes that are important in negotiations with governments, industrial entrepreneurs and others who have an interest in a proposed development, but are not from the area. These are to:

- identify hazards at an early stage and plan how to reduce these if the project goes ahead;
- increase protection of the existing environment, its peoples, flora and fauna, including their health and the traditional lifestyles and cultures that the people represent;
- ensure sustainable use of natural resources;
- develop better designs for the proposed project;
- reduce project costs and delays by thinking about potential problems before they arise and devising possible solutions for them;
- increase the accountability of government and industrial entrepreneurs to those affected by their decisions; and
- encourage direct participation by those affected by the decisions, especially First Nations peoples.

The consultation process used to collect information and ascertain people's views on a proposed project is important. First Nations acknowledge the uncertainty that exists around these developments – information may be incomplete, mitigation measures may be difficult to implement, predictions are unreliable because new information may be acquired at any time and this may change the basis of the initial decisions. Thus, an environmental assessment has to be seen as incomplete and constantly re-examined in light of new evidence. This is particularly relevant when considering the impact of an installation on the surrounding area, when access roads, temporary accommodation for workers and resettlement of displaced populations (if any) have to be considered. The FNEATWG toolkit was used to reach a decision in the case study that follows.

Case study

BC Hydro proposed a gas-fired electricity plant in Nanaimo, Canada, to address a projected growth in electricity demand on Vancouver Island of 1.6 per cent per annum. Its construction was expected to cost CDN$710 million and meant that some ageing components of the electrical transmission system currently in use could be replaced at the same time. As the electricity would be locally produced energy, the project would reduce reliance on the mainland for energy needs on Vancouver Island. The energy to be supplied by the plant in Nanaimo was to be supplemented with electricity that would be generated by a new facility that was to be built in Port Alberni, also on Vancouver Island.

The proposal was considered by the Snuneymuxw First Nations peoples on whose land the Nanaimo project would have been built. When they conducted their own analysis and environmental assessment, they were not convinced by the arguments and data in the original proposal, and raised concerns and questions about the data provided. These covered projected rises in energy demand; the lack of transparency in the data collection and analysis of materials provided; the inadequacy of the way in which data were interpreted; poor use of the forecasting model to evaluate the figures for past, actual and predicted energy usage; and absence of alternative possible solutions.

The Snuneymuxw consulted with their peoples and sought experts of their own choosing, to advise them and help evaluate the projected proposals. At the end of this process, they decided that there was insufficient evidence to justify why gas-powered facilities that required a new pipeline under the Georgia Strait was the proposing company's preferred solution. The projected rises in population and increases in energy consumption used were not valid as the firm had projected higher usage in the past that had not materialized. They felt that the proposed arrangements were inadequate for any potential shortfall that might arise; and that ageing parts of the existing infrastructure could be replaced by other means with less financial cost and damage to the environment. The Snuneymuxw refused to approve the project. The BC Utilities Commission also rejected the application and asked the company to 'properly conduct an identification of alternatives for power generation on Vancouver Island' (FNEATWG, undated: 9).

The experience of the Snuneymuxw shows that environmental assessments are useful in carrying out a thorough investigation and analysis of proposals for development that can cause irreparable harm to the environment and damage people's quality of life, culture and livelihoods. It also shows how important it is to consult effectively with people and bring in expertise, especially that of a scientific kind, to enable them to assess fairly and competently any information that companies provide in what prove to be contentious situations. Throughout this process, collective action and collaborative learning have been at the centre of the Snuneymuxw's actions. Their consultations have resulted in decisions that they can defend in public arenas with others, and, obviously, those in power have listened carefully to their arguments and supported their views, as the BC Utilities Commission did in this case study. Indigenous social workers can facilitate and coordinate these processes, and they have been involved in environmental decision-making processes as activists, professionals and members of specific First Nations bands. These practitioners can also help First Nations peoples access a range of professionals who can also provide additional expert advice to a specific community. Moreover, the First Nations peoples have indicated their willingness to share their learning and knowledge by making the Environmental Framework of Assessment that they have devised freely available to those who wish to use it. This stands in sharp contrast to the restricted access given by those who control information flows through copyright agreements.

Conclusions

Indigenous ways of being, perceiving and acting in the world are closely linked to their notions of spirituality and the close and valued connections they have between people, other living things, inanimate objects and the rest of the ecosphere, including its water, air and land. They seek a symbiotic relationship in their dealings with the natural world, not the exploitative one that often features in Western modernity's models of industrialization. The failure of big business to take care of the environment properly has been a major concern of environmentalists across the globe, including social workers (Besthorn, 2011). This view has been adopted because big business is now seen as a major contributor to the greenhouse gas emissions that threaten the world, and there is only limited evidence that large corporations are making substantial inroads into remedying the problems that their activities have created by degrading the environment and jeopardizing peoples' health.

Non-indigenous social workers have much to learn from indigenous worldviews and holistic approaches to life. They can incorporate these in their practice to support people better in local communities and both meet their aspirations for an improved quality of life and care for the environment. The obligation to care for the environment and be cared by it in return lies at the heart of sustainable development, regardless of where it is being carried out and by whom.

10

Conclusions: Green Social Work

Introduction

The social, fiscal, environmental crises affecting social work practice early in the twenty-first century call for courage and innovation in facing the challenges that these present for ordinary people, academics, practitioners, policy-makers and students. Developing new theories and paradigms for practice is an imperative that cannot be ignored if the well-being of humanity and the planet are to be affirmed and realized. Sharing the Earth's human, social and physical resources within a green framework is best achieved through interdisciplinary, transnational and egalitarian partnerships among multiple stakeholders who respect each others' differences whilst searching for commonalities that will enable every living thing and the physical environment to be cared for and flourish. This is a task that social workers, with their holistic vision and concern for those who are marginalized, disempowered and without voice at decision-making tables, can engage with and promote in the various ways discussed in this book.

I conclude my writing by considering how the insights explored in the previous chapters can help to develop a holistic model of social work practice that acknowledges interdependence and solidarity among the world's peoples, flora, fauna; uses material resources sustainably and equitably; and develops sustainable lifestyles to preserve and enhance the well-being of all peoples and the planet. I have called the model articulated in this book, *Green Social Work*. It involves social workers working closely with people in their communities in and through their everyday life practices to:

- respect all living things alongside their socio-cultural and physical environments;
- develop empowering and sustainable relationships between people and their environments;
- advocate for the importance of embedding the social in all economic activities including those aimed at eradicating poverty;
- question the relevance of an industrial model of development that relies on over-urbanization and over-consumption as the basis for social progress; and
- promote social and environmental justice.

Achieving these goals has social workers engaging in action at the local, national, regional and international levels and using the organizations that they have formed to transform existing policies that favour the unequal distribution of power, social resources, services and commodities as well as focus on the protection of the Earth's physical bounty, including its flora and fauna alongside its peoples.

Developing Holistic, Sustainable Practices

Green social work engages with the environment as a discursive terrain in which people live out their narratives of place and space. These narratives express their meaning and purpose in the world and define their relationships with other peoples, the flora, fauna and material existences about them. Green social work is underpinned by a moral and ethical approach that is rooted in the spiritual notion that there are relationships between people, other living entities and the physical realm. It also endorses the provision of publicly funded social services available to all at the point of need, and the right to care for others and be cared by them. The state becomes the guarantor of these rights for all its residents. I recognize the contested nature of both the epistemological and ontological nature of these statements and do not claim that one size of development fits all. So, I do not provide a monolithic model of practice in *Green Social Work,* although I offer guidelines that are consistent with a moral and ethical standpoint that advocates for an equitable distribution of the Earth's resources, a collective pooling of risks and benefits, and a duty of care towards the world and all living and material things in order to enjoy being cared for through the bounty that nature provides. The ideologues of modernity have taken for granted the availability of the Earth's resources for appropriation and exploitation by whoever could do so, namely those with money. It is this indifference to the Earth's physical reality and the treatment of the world's resources, including the minerals, flora, fauna and peoples within it, as means to an end that has undermined the well-being of humans and the biosphere. Current models of industrialization, private ownership of productive factors including land, water and other physical resources provided by the Earth by virtue of its existence, and organizing these for the profit of a few individuals sit uneasily within an egalitarian green framework. This form of ownership, generally described as a possessive individualistic neoliberal capitalism, provides only one means whereby resources can be allocated to individuals and/or groups: the market-place. This model is unfair because it ensures that risks are spread out among the many to privilege the few. In trying to transcend the limitations of neoliberal capitalism through an inclusive egalitarian framework predicated on social and environmental justice, green social workers adopt an explicitly political stance, rather than the covert one inherent in neoliberal ideology in which the current political regime is seen as non-political or neutral. They also challenge those in politics, the professions, the social and physical sciences and the Earth's inhabitants in general to work together to co-produce alternative ways of organizing social relations that are fair, egalitarian and socially and environmentally just.

Green Social Work also promotes the idea that practice is locality and culturally specific, while at the same time espousing the view that there are important concerns that are embedded in the interdependencies that exist between human beings and the Earth's flora and fauna that are relevant across the world and that must be incorporated into local practices if the well-being of all is to be assured. So, a key task for green social workers is to engage people in discussions about these interdependencies, to understand their implications for local behaviours and to work collaboratively with others to ensure their realization at all levels – from the local to the global and back. Hence, green social work practice both addresses the uniqueness and specific nature of the local, and highlights its interaction with global issues, policies and practices that involve people and their relationship with the Earth's flora, fauna and physical environments. The 'one size fits all' model of practice, typical of many paradigms that have been popular in the social work profession to the present, has been rooted primarily in the parochialism of local experiences, rather than this expansive view of the local and the interpenetration of the local with the global and the global with the local.

Moreover, the profession has disregarded interdependencies and the impact of other peoples and living things upon the local while simultaneously suggesting that the local encompassed experiences beyond it by assuming that these were alike, or that these should be if they were not, in an imposed universality of existence. This pattern was reproduced even when it was clearly inappropriate, and at times destructive, as indigenous writers have indicated has applied to them over the years. Sustainable lifestyles were central to indigenous ways of living. Indigenous practices have offered and continue to offer understandings that can assist in the development of new paradigms for practice today. This occurs through their being rooted in the local while at the same time reaching out to the universal by transcending space and time and linking current-day practices with those of their ancestors and their successors in a dynamic that respects bio-diversity, preserves life-enhancing traditions, develops resilience and adapts to change and threats, described by Bruyere (2010) as 'Picking up what was left by the trail'. The past, present and future are interdependent, but the links between them have to be maintained to ensure survival across the time and space dimensions. Indigenous people also have a close relationship between peoples, the spiritual realm and the living and physical environments. The dynamic relationships between the different parts of green social work are indicated in figure 10.1. Moreover, as people negotiate their realities, it allows for change.

The hegemonic neoliberal orientation of contemporary societies presents green social work with one of its major challenges. That is, ensuring that green social work plays a key role in eradicating poverty while caring for and protecting the environment within a framework of social and environmental justice, human rights, active citizenship and a critique of neoliberal capitalist modes of production, distribution and consumption. Social workers have to tackle poverty because it is the condition in which the majority of those with whom they work live. Poverty is a disaster in its own right because it affects so many people. And it is one that needs urgent attention. Eradicating poverty will be complicated and contested, not least because it is a structural and highly political issue. Moreover, there are many vested interests involved in maintaining the world order as it is. The allocation of power and resources, including access to public resources, services and goods, is unequal, and the history of the welfare state in continuing to marginalize poor people has meant that middle-class claimants get more out of the benefits the welfare state offers. For example, public subsidies in agriculture and industry encourage inefficient and profligate use of resources in both the Global North and the Global South. Removing public subsidies from agricultural and industrial production and consumption could be helpful in reducing the wasteful and inefficient use of resources (Guzmán et al., 2009). However, this leaves intact the problem of facilitating access to resources for those who lack the necessary funds. Meeting their needs is a key concern of green social workers and provides the impetus for developing alternative forms of production, distribution and consumption.

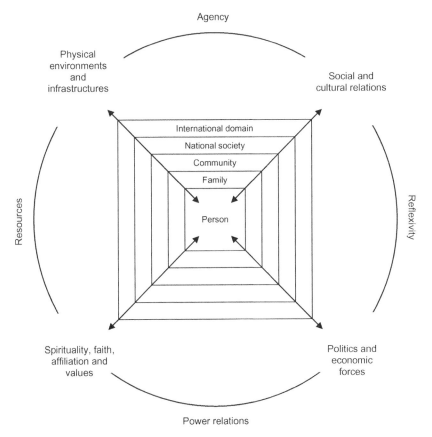

Source: adapted from Dominelli (2002)
Figure 10.1   Holistic practice chart

Poverty eradication cannot happen in a context that *assumes* that all consumers have equal access to goods and services, including water, food and energy. Equality has to be worked for to be realized. The market-based free-for-all has to end. Those without adequate incomes cannot be left as non-players in the market-place as long as that remains the main mechanism for distributing goods and services. They will have to access the world's resources, either as entitlement and as of right, or in the form of a decent guaranteed income for all, or in terms of every person having a well-paid job that pays sufficient wages for the purchase of the basic necessities of life, historically defined. Benefits and salaries or wages would have to be set at a level that would be sufficient for each person to have decent standards of living, including clean drinking water and sanitation services. Paid work should not consume the whole of a person's life or be the only factor considered in developing alternative economic paradigms for meeting the interdependent needs of people and the environment. Recreation, leisure and family life should be included in the equation. These entitlements are based on the urgency of realizing human rights, social and environmental justice and active citizenship for marginalized groups, especially women and children. Achieving these will require a more equitable sharing of the Earth's bounty for all, and new paradigms of production and consumption. Social workers can play key roles in advocating for such changes, undertaking research that demonstrates the suffering of poor and marginalized peoples across the globe in the absence of alternatives to market-based allocations of goods and services; promoting robust resilience in communities; and developing alternative models of production and consumption.

Finding alternatives to the excesses of industrialization following capitalist models of development is essential in safeguarding life on Earth. Hyper-urbanization and agribusiness have provided many opportunities for some people, but they have also excluded around 50 per cent of the world's population who live in poverty, have very limited prospects and inhabit degraded environments where the depletion of the Earth's resources has already occurred. Some of the activities green social workers would focus upon include working with people to articulate alternative visions based on realizing their dreams for a more sustainable, life-enhancing world. They would do this at individual and group levels, in communities, locally, nationally and internationally, helping people to mobilize collectively, and take action to achieve their ambitions. This requires a respectful partnership between nature and society. The two are part of one whole project of life-affirming replenishment. Those adhering to the deep ecology perspective have expressed these goals in their claim that 'social work practice needs to address the problems that arise from excessive and destructive human interference with nature' (Van Wormer & Besthorn, 2011: 249). Thus, replacing this destructive relationship with a constructive sustainable one becomes an important task for green social workers.

Green social work has considerable implications for contemporary humanitarian aid during disaster relief because its current forms do not focus specifically on environmental and ecological issues, even though aid is usually connected to these as assistance is given within the context of degraded environments aggravated by disasters. When integrating green perspectives into disaster relief agendas, the separation and fragmentation between experts and people has to be overcome. Social workers can bring these disparate groups together and coordinate activities so that they begin to work cooperatively as a unit, and help local residents to rebuild their lives and environments through active, engaged and empowering processes. The co-production of knowledge and solutions to problems will be integral to changing current practices whereby each professional grouping works separately (Dominelli, in press).

Urbanization developed because business preferred to centralize its activities to gain economies of scale and lower the costs of production. Whether urbanization is an appropriate model to use, given that moving workers across geographic terrain, either as daily commuters or as migrants, adds substantially to energy consumption, environmental degradation and people's ill-health, is debatable. Urbanization has also contributed to depopulating rural areas. Developing holistic systems that capture the complex dynamics at play in such situations and completing assessments that are fit for purpose and inclusive of all considerations, not just economic costs, are essential. Social workers, as the people charged with understanding and intervening in people's daily lives on a holistic basis, can assist in the process of researching and determining these holistic costs by working alongside other experts and the residents and communities directly affected to find solutions that will care for both peoples and their environments.

Working for a Sustainable, Interdependent and Healthy Planet that Nurtures All Peoples and Their Environments

Developing a sustainable, nurturing planet for all peoples on Earth to benefit from equally requires social workers to think differently about people's relationships between themselves and their environments – social, emotional, spiritual and physical. Rosenhek (2006: 19) put this succinctly and simply when he said: 'the deep ecology movement reminds us that we are from the Earth, of the Earth and not separate from it'. In other words, the relationship between people and the Earth is not divisible, nor can the well-being of one be separated from that of the other. They are two parts of one whole, each dependent on the other for a long, sustainable and healthy existence.

Social work interventions that address crises and disasters of whatever kind involve assessments of risk that aim to: minimize the objective possibility of such events occurring; tackle the subjective fear that a calamity will strike; and develop resilience that will address areas of vulnerability. Social workers endeavouring to reduce risks in the multiple environments inhabited by people will be engaging in green social work with individuals, groups and communities locally, nationally and inter-nationally. They will also cover the policy arena and the academy in teaching, research and practice. Their involvement will endeavour to get people to:

- reduce the potential for jeopardizing health;
- modify identified hazards;
- decrease vulnerabilities and increase robust resilience;
- improve preparedness for future disasters; and
- work on disaster mitigation and prevention.

As practitioners, social workers will adopt a number of roles while doing this work, as shown in the list below. Social workers act as:

- facilitators
- coordinators
- community mobilizers
- resource mobilizers and coordinators
- negotiators
- mediators
- consultants
- advocates
- educators
- trainers
- cultural interpreters
- psycho-social therapists
- scientific translators.
  (Adapted from Dominelli, 2010a)

The intervention processes that they can engage in are those that social workers are already familiar with, regardless of area of practice, and are listed below. These processes are part of good practice, but they should not be anticipated as linear or unproblematic in their implementation, especially if the communication and transportation networks are inoper-able, as occurs regularly in disaster situations. The intervention processes the social workers utilize in helping relationships are:

- making initial contact
- engaging local people throughout the processes of intervention
- acknowledging local people's existing knowledge and skills
- co-producing new knowledge and skills relevant to the particular community and habitat(s)
- thinking holistically
- assessing available information and drawing upon the latest research data
- identifying areas of work
- formulating work plans
- drawing up contracts for the work to be undertaken
- initiating the agreed action(s)
- constantly evaluating the results
- finding sources of support, including material resources, social networks, expertise and allies.

Paying attention to the processes of intervention is important because these focus on how the work is done and provide the context whereby people judge their experiences of interventions as inclusive and empowering or as expert-led and disempowering. Social workers undertake the following activities to develop resilience among people and communities despite practising in complex disasters and under difficult environmental conditions.

- Raising awareness, including engaging in discussions about different disaster scenarios and their outcomes, before, while and after an emergency situation arises.
- Holding dialogues about policies with policy-makers, practitioners and service users to change policy and encourage the sharing of resources and green technologies across the world. Green technologies encompass preventative technologies to safeguard people and the physical environment, e.g., early warning systems for tsunamis, renewable energy sources.
- Lobbying for preventative measures locally, nationally and internationally.
- Mobilizing individuals, communities and resources and developing resilient responses before, while and after disaster strikes.
- Researching issues and problems to formulate new ways of looking at the world, and of living and being within it.
- Training people in how to value and respect others, including those who hold different views from them. This does not mean having to agree with everyone, but disagreements should leave persons feeling

that they have been treated with dignity and that their arguments have been heard even if, at the end of the process, they simply agree to differ.

- Observing a duty of care towards the environment and working to protect it.
- Assisting in creating egalitarian, sustainable models of development.
- Developing curricula for mainstreaming green social work in qualifying and post-qualifying social work programmes and other relevant disciplines.

Green social workers operating in circumstances aimed at empowering people and realizing their human rights and citizenship entitlements have the duty of acting morally and ethically. The moral and ethical principles that guide this work are the following.

- Human rights and dignity at individual and group levels
- Interdependence, reciprocity, mutuality and solidarity
- Social justice
- Environmental justice
- Celebrating cultural diversity
- Preserving biological diversity
- Peace

These principles enshrine values that go beyond green social work and a number of these – especially the first three and the last one – would be recognized by practitioners working in anti-oppressive ways.

Building Capacity in Communities

Communities are contested entities because a range of stakeholders is involved in determining their boundaries and criteria for inclusion and exclusion. Some parties to the social exchanges involved in community creation may be more interested in regulating these once they are in place. Others may want very loose rules concerning membership and actions. In any case, the definitions of community are varied, although the literature tends to break these down into three main typologies: geography, identity and interests. Communities, according to Delanty (2003: 187), are 'created through communicative practices and normative concepts' rather than social structures or cultural values. In holistic practice, social structures, cultural values and daily routines become important elements in community formation because culture and structure provide some of the content of communicative practices and conceptual norms. Moreover, community formation is an unfinished process involving constant change and adaptation. And, as existing communities form and reform themselves, their transformation can occur along any point of the political spectrum from left to right (or right to left). Radicalism is not the prerogative of the right. Community boundaries tend to be porous and people's sense of belonging to one may be fluid because they can be members of more than one community simultaneously, and communities change over time. Communities are relational and interactive fields that are formed through a variety of discourses. These discourses can be authored by a number of stakeholders, including residents, outside experts, politicians, practitioners, and through state legislation, policies, organizational routines and everyday life practices. A good starting point for conceptualizing the process of forming and developing communities is Emejulu's (2011: 230) definition of community development as 'a political and social process of collective education and action to achieve self-determination and social justice for marginalized groups'.

A commitment to democracy and a set of democratic practices are important in community formation processes because these provide the means for including all voices in discussions about community. This involves organizing 'democratic spaces' for learning and action that consider how people are configured; how decisions are made and by whom; and enhance their capacity to act collectively and in solidarity with others (Emejulu, 2011). To operate in such an environment, people's agency and capacity to determine and initiate action on their behalf has to be to the fore, and they need to invent a process that enables them to value dissenting voices by hearing and engaging with, rather than marginalizing, them. The trick here is not to dominate people or replace one set of unhelpful rulers with another. However, given that we live in a world in which binary categorizations of people are the norm, operating in effective 'democratic spaces' will not be easy. To begin with, marginalization is a by-product of hegemonic social relations. Marginality is the flip side of dominance. Finding alternatives to hegemonic realities and undertaking action to replace these with egalitarian ones is a huge task for the egalitarian democratic practitioner who wants to work in a holistic and inclusive manner, because these activities have to be conceptualized and negotiated together with all the people involved. In such circumstances, the outcomes can be very uncertain.

If these are ecologically oriented, geographic and spatial communities, they have a solid basis to them. Working alongside interest communities, people can consider both social and environmental considerations more readily than those based on identity, unless this is associated with space or place, because concrete notions of locality are easier for individuals to relate to when thinking about abstract concepts like sustainability.

Resilience is an important component of community development because this characteristic is formed through the collective negotiation of the norms that bind together people who may be very different from each other through discourses about 'diversity, social justice and peaceful coexistence' (Kuecker et al., 2011: 253). Moreover, their search for common ground, which they establish through their negotiations, will help allay uncertainties about scarcities that undermine the sense of belonging to a community in a number of contemporary discourses, e.g., those about migrants settling in already stressed locations among disadvantaged people and drawing upon stretched local resources, especially welfare provisions, because preparations for their arrival have been either inadequate or non-existent. Community workers intervening in such situations can work to enhance the sense of belonging to a particular place among all residents by creating dynamic relationships that can cope with changing a community's sense of itself and affirming it as an inclusive space where problems are shared and solved collectively. This requires reducing local tensions and eliminating potential conflicts by dialoguing across differences, reducing social isolation and fragmentation between groups and undertaking agreed inclusive collective action around specific problems. An approach like this is essential in addressing the needs of climate change migrants in particular, because they are not covered by current international conventions (McInerney-Lankford et al., 2011). In carrying out inclusive projects, community workers ensure that they do not render existing cleavages invisible, but seek to address these and reduce any underlying tensions these divisions create, especially in multicultural communities that have a plurality of cultures and traditions. Establishing full and active citizenship rights, forming alliances and working with allies across many different types of social divides become crucial parts of this process, as does caring for and about an ecosystem that would sustain all community members.

The struggle to maintain community integrity may be more poignant in communities in the Global South, where residents might be forcibly moved during armed conflicts, or to make way for 'developments' that uproot poor people for luxury housing, corporate headquarters and extractive industries that degrade the local environment and jeopardize health among workers and their families. Such ventures often encounter resistance as occurred in Junín in Ecuador (Kuecker et al., 2011). I describe this situation in the box below.

Case study

Junín is an agrarian community in northwest Ecuador that has been ear-
marked for extensive mining developments, especially those involving copper
since the 1990s. Several corporate proposals sought open-pit mines that
would have threatened to degrade tropical Andean forests, destroy diverse
and rare wildlife, poison the water supply and undermine people's quality of
life, including through the relocation of residents living in the areas where
the proposed mines were to be located. The Junín struggle to resist such
projects has been strong and continuous, involving the formation of a grassroots movement including DECOIN (Defensa y Conservación Ecológica de
Intag) and Acción Ecológica. These became part of a coalition that comprised
local residents, ecological activists, liberation theologists, scientists, overseas
supporters and civil society organizations aimed at protecting the environment and the livelihoods of indigenous peoples residing in the area. The
community also passed legislation that enabled it to become an 'ecological
canton' that required mining companies to consult with local populations
through participative planning. Eventually, open-pit mines were banned by
law. Additionally, opposition to the plans of multinational corporations
included government ministers and the mayor of Cotacachi criticizing their
proposals on several occasions. Significant bodies such as the country's
Comisión de los Derechos Humanos (CEDHU, or the Ecumenical Commission on Human Rights), the United Nations and Amnesty International also
supported the local population's claims. Although the UN's Global Compact
on Mining placed an onus on global corporations to uphold the human rights
of local inhabitants, these were generally disregarded. The protracted controversies between local people and the mining interests have become violent at
times. Still, the people continued to mobilize and insist on their right to
develop in their own way. This has resulted in alternative forms of income-
generation schemes such as growing organic coffee, ecotourism and selling
handicrafts to offset potential funds that could have flowed into the region
from extractive industries. Although the mining companies are not currently
operating, Kuecker et al. (2011) suggest that they have not given up. They
are simply waiting for a further opportunity to re-assert their wishes.

The struggle of the indigenous peoples of Junín to maintain a way of
life that is in harmony with the ecosystem offers important lessons for
green social workers. It demonstrated the importance of believing in the
'right to have rights' – an essential ingredient of active citizenship – and
having strong solidarity in resisting the development of a rural community for industrial purposes. Dialogues across differences successfully
galvanized a marginalized agrarian society to take collective action that
protected ancestral spaces. In the process, the expanded development of

existing bonds of mutual (inter)dependence that stretched to supporters from other countries, building on community strengths, maintaining a view of the wider picture, and enhancing social networks, were significant in sustaining this particular social action over time. The processes of mobilization reinforced feelings of solidarity and reciprocity already present, and promoted a combined sense of overall purpose in the face of specific threats. The Junín initiative also drew on community labour provided through a traditional custom called *minga* (Kuecker et al., 2011) and enabled its members to support each other in defending their community interests. Their situation also highlights how local boundaries can be penetrated by the global when multinationals seek to impose their will upon the territory they target, and illustrates how local resistance impacts on the global when such companies are obliged to change their plans in response to local mobilizations. This example also poses questions about the role of the state and whether it sides with its citizens or the corporate elite. Such queries exercise the mind of green social workers, given their critique of neoliberalism and the role of global corporations in exploiting both people and the Earth's resources for the benefit of the few. Redressing the imbalances in power relations between those representing international finance and local people will not be easy.

Conclusions

Green social work has set an ambitious but worthwhile agenda for the social work profession. Drawing on the right of peoples to have rights to care for each other and the biosphere and be cared by them in return, it calls for the constant development of new concepts, theories and models for practice that engage with a range of stakeholders to develop locality-specific and culturally relevant responses to the major problems of our time – poverty, environmental degradation, unsustainable forms of development and industrialization, extinction of species, resource shortages – and address global interdependencies. Green social workers have attempted to find solutions that are developed in localities, but have resonance at national, regional and international levels because they speak to social problems that affect the human condition across the world and which must be resolved by drawing on the contributions of all of the Earth's inhabitants. Green social work provides an optimistic thread that links past, present and future together by: drawing on solidarity, equality; working in partnership with nature to address structural inequalities that have stunted human potential for the many while destroying much of the physical environment in the process; and highlighting our dependency on each other and the Earth's flora, fauna and natural resources. In this scenario, the relationship between production, reproduction and consumption need not be the antagonistic one depicted in Marxist writings, although there may be conflicts of interest and differentiated privileging in social organization as currently practised. But this need not be so. As a respondent in 'Internationalizing Institutional and Professional Practices' said, 'Good partnerships between people offer hopes for a better present and future' (Dominelli, in press). People can devise alternative economic models that: encompass all the Earth's inhabitants and their rich and diverse cultures; share resources equitably; and start off with the social – the realm that people inhabit – alongside the physical – where the flora, fauna and physical resources reside. All are located as equal partners in an equation that aims to provide a decent quality of life for all living things, sustain the material environment and ensure the existence of all these in both the present and the future. Social workers have a pivotal role to play in this through the mobilization of communities in empowering processes that link the local to the global and the global to the local, in one world that has to be shared for the benefit of all. At the end of the day, green social work offers a model for good social work practice.

Bibliography

Abdisaid, M. (2008) *The Al-Shabaab Al-Mujahidin: A Profile of the First Somali Terrorist Organisation*. Berlin: Institut für Strategie Politik Sicherheits und Wirtschaftsberatung (ISPSW).

Action Aid (2011) *Building for the Future*. London: Action Aid.

Adger, N. and Vincent, K. (2005) 'Uncertainty in Adaptive Capacity', *Geoscience*, 337: 399–410.

Agenda 21 Plus (2004) *Agenda 21 for Culture*. Available at www.agenda21culture.net and www.barcelona2004.org. Both accessed 2 August 2011.

Agyeman, J. and Evans, B. (2004) '"Just Sustainability": The Emerging Discourse of Environmental Justice in Britain?' *Geographical Journal*, 170(2): 155–64.

Agyeman, J., Cole, P., Haluza-Delay, R. and O'Riley, P. (eds.) (2009) *Speaking for Ourselves: Environmental Justice in Canada*. Vancouver: University of British Columbia Press.

Aiken, M., Cairns, B. and Thake, S. (2008) *Community Ownership and Management of Assets*. York: Joseph Rowntree Foundation.

Allan, J. P. and Scruggs, L. (2004) 'Political Partisanship and Welfare State Reform in Advanced Industrial Societies', *American Journal of Political Sciences*, 48(3): 496–512.

Allouche, J. (2007) 'The Governance of Central Asian Waters: National Interests versus Regional Cooperation', *Disarmament Forum*, 4: 44–55.

Alston, M. (2002) 'From Local to Global: Making Social Policy More Effective for Rural Community Capacity Building', *Australian Social Work*, 55(3): 214–26.

Amin, A., Cameron, A. and Hudson, R. (2002) *Placing the Social Economy*. London: Routledge.

Anand, P. B. (2004) *Water and Identity: An Analysis of the Cauvery River Water Dispute*. BCID, Research Paper 3. Bradford: Bradford University.

Angelsen, A. (ed.) (2008) *Moving Ahead with REDD: Issues, Options and Implications*. Bogor Borat: CIFOR.

Anishchuk, A. (2010) 'Fires Could Stir Up Chernobyl Radiation', *The Vancouver Sun*, 12 August, p. B4.

Arblaster, L., Conway, J., Foreman, A. and Hawkin, M. (1996) *Inter-agency Working for Housing, Health and Social Care Needs of People in General Needs Housing*. York: Joseph Rowntree Foundation.

Austin, M. (1983) 'The Flexner Myth and the History of Social Work', *Social Services Review*, 57(3): 357–77.

Averchenkova, A. (2010) *The Outcomes of Copenhagen: The Negotiations and The Accord*. New York: UNDP.

Bachir, H. (2008) *Tears of the Desert: A Memoir of Survival in Darfur*. London: One World, Ballantine.

Bader, M. (2009) 'The Government's Carbon Off-Set Scheme Is a Scam', *The Vancouver Sun*, 2 December.

Balk, D., Montgomery, R., McGranahan, G., Kim, D., Mara, V., Todd, M., Buettner, T. and Doréelien, A. (2009) 'Mapping Urban Settlements and the Risks of Climate Change in Africa, Asia and South America' in J. Guzmán, G. Martine, G. McGranahan, D. Schensul and C. Tacoli (eds.) *Population Dynamics and Climate Change*. New York: UNFPA/IIED.

Barker, H. (1986) 'Recapturing Sisterhood: A Critical Look at "Process" in Feminist Organisations and Community Action', *Critical Social Policy*, 16(3): 80–90.

BBC News (2010) 'Short History of Immigration', *BBC News Online*. Available at http://news.bbc.co.uk/hi/english/static/in_depth/uk/2002/race/short_history_of_immigration.stm. Accessed 20 April 2010.

BCWWA (British Columbia Water and Waste Association) (2011) 'Drinking Water Week', *The Vancouver Sun*, 27 April, pp. A8–A9.

Benzie, M., Harvey, A., Burningham, K., Hodgson, N. and Siddiqi, A. (2011) *Vulnerability to Heatwaves and Drought: Case Studies of Adaptation to Climate Change in South-West England*. York: Joseph Rowntree Foundation.

Bertazzi, P. (1991) 'Long-term Effects of Chemical Disasters: Lessons and Results from Seveso', *The Science of the Total Environment*, 106(1–2): 5–20.

Besthorn, F. (2004) 'Globalized Consumer Culture: Its Implications for Social Justice and Practice Teaching in Social Work', *The Journal of Practice Teaching in Health and Social Work*, 5(3): 20–39.

Besthorn, F. (2008) 'Environment and Social Work Practice' in *Encyclopaedia of Social Work*. Oxford: Oxford University Press. Second edition.

Besthorn, F. (2011) 'The Deep Ecology's Contribution to Social Work: A Ten Year Perspective', *International Journal of Social Welfare*, on-line version, 9 December.

Besthorn, F. and Meyer, E. (2010) 'Internally Displaced Persons, Broadening Social Work's Helping Imperative', *Critical Social Work*. Available at www.uwindsor.ca/criticalsocialwork/2010-volume-11-no-3. Accessed 3 August 2011.

Biermann, F. and Boas, I. (2010) 'Preparing for a Warmer World: Towards a Global Governance System to Protect Climate Refugees', *Global Environmental Politics*, 10(1): 60–88.

Bishop, J. (1996) *Becoming an Ally*. Halifax, NS: Fernwood Publishing.

Bishop, M. and Green, M. (2008) *Philanthrocapitalism: How the Rich Can Save the World*. London: Bloomsbury Press.

Bissio, R. (2011) *After the Fall: Time for a New Deal, Social Watch Report 2010*. Available at www.socialwatch.org. Accessed 20 October 2011.

Blanchard, B. (2011) 'Higher Sea Levels Add to China's Disasters', *The Vancouver Sun*, 25 April, p. B5.

Bocquier, P. (2008) 'Forecast No. 8, Urbanization to Hit 60 Per Cent by 2030 Causing More Epidemics and Environmental Problems, *The Futurist*. Available at www.britannica.com/blogs/2008/12/forecast-8-urbanization-to-hit-60-by-2030/. Accessed 29 April 2011.

BoE (Bank of England) (2003) *Financing for Social Enterprises: A Special Report by the Bank of England*. London: BoE.

Bolger, S., Corrigan, P., Dorking, J. and Frost, N. (1981) *Towards a Socialist Welfare Practice*. London: Macmillan.

Bolz, D. (2009) 'Endangered Site: The City of Hasankeyf, Turkey', *The Smithsonian*. Available at www.smithsonianmag.com/travel/Endangered-Cultural-Treasures-The-City-of-Hasankeyf-Turkey.html. Accessed on 2 August 2011.

Bornstein, D. (1996) *The Price of a Dream: The Story of the Grameen Bank and the Idea that It Is*. New York: Simon and Schuster.

Borrell, J., Lane, S. and Fraser, S. (2010) 'Integrating Environmental Issues into Social Work Practice: Lessons Learnt from Domestic Energy Auditing', *Australian Social Work*, 63(3): 315–28.

Boswell, R. (2010) 'Massive Iceberg Threatens Ships, Oil Rigs', *The Vancouver Sun*, 11 Aug, p. B2.

Boyden, J., and Mann, G. (2005) 'Children's Risk, Resilience, and Coping in Extreme Situations' in M. Ungar (ed.) *Handbook for Working with Children and Youth: Pathways to Resilience across Cultures and Contexts*. London: Sage Publications.

Brahic, C. (2007) 'Unsustainable Development "Puts Humanity at Risk"', *New Scientist*, 15: 17, 25 October. Also available at www.newscientist.com/article/dn12834-unsustainable-development-puts-humanity-at-risk.html?full=true&print=true. Accessed 20 January 2011.

Brender, N., Cappe, M. and Golden, A. (2007) *Mission Possible: Successful Canadian Cities and Sustainable Prosperity for Canada*. Ottawa: The Conference Board of Canada.

Bronfenbrenner, U. (1977) 'Toward an Experimental Ecology of Human Development', *American Psychologist*, July: 513–31.

Brown, O. (2008) 'The Numbers Game', *Forced Migration Review*, 31(October): 8–9.

Brown, P. and Garver, G. (2009) *Right Relationship: Building a Whole Earth Economy*. San Francisco: Berrett-Koehler Publishers Inc.

Brox, J. (2008) 'Infrastructure Investment: The Foundation of Canadian Competitiveness', *Policy Matters: Institute for Research in Public Policy*, 9(2/August): 1–48.

Brundtland, G. (1987) *The World Commission on the Environment and Development: Our Common Future*. Oxford: Oxford University Press.

Bruyere, G. (2010) 'Picking up What was Left by the Trail: The Emerging Spirit of Aboriginal Education in Canada' in M. Gray, J. Coates and M. Yellow-Bird (eds.) *Indigenous Social Work Around the World: Towards Culturally Relevant Education and Practice*. Aldershot: Ashgate.

Bullard, R. (2000) [1990] *Dumping in Dixie: Race, Class, and Environmental Quality*. Boulder, CO: Westview Press. Third edition.

Burkett, I. (2007) 'Globalised Microfinance: Economic Empowerment or Just Debt?' in L. Dominelli (ed.) *Revitalising Communities in a Globalising World*. Aldershot: Ashgate.

Byrne, D. (2005) *Social Exclusion*. Maidenhead: McGraw-Hill.

Cabinet Office (2008) *The Pitt Review: Learning Lessons from the 2007 Floods*. London: Cabinet Office.

CAG Consultants (2009) *The Differential Social Impacts of Climate Change in the UK*. Edinburgh: SNIFFER (Scotland and Northern Ireland Forum for Environmental Research).

Cahill, M. and Fitzpatrick, T. (eds.) (2002) *Environmental Issues and Social Welfare: Broadening Perspectives in Social Policy*. Oxford: Blackwell Publishers Ltd.

Callahan, M., Dominelli, L., Rutman, D. and Strega, S. (2002) 'Undeserving Mothers? Practitioners' Experiences Working with Young Mothers in/from Care', *Child and Family Social Work*, 7: 149–59.

Cameron, D. and Clegg, N. (2010) *Building the Big Society*. Available at www.cabinetoffice.gov.uk/news/building-big-society. Accessed 26 April 2011.

Carrington, D. and Vidal, J. (2011) 'Global Food System Must Be Transformed on "Industrial Revolution Scale"', *The Guardian*, 24 January. Available at www.guardian.co.uk/environment/2011/jan/24/global-food-system-report. Accessed on 28 January 2011.

Carruthers, D. and Rodriguez, P. (2009) 'Mapuche Protest, Environmental Conflict and Social Movement Linkage in Chile', *Third World Quarterly*, 30(4): 743–60.

Carson, R. (1962) *The Silent Spring*. New York: Houghton Mifflin Co.

Catney, P. and Doyle, T. (2011) 'The Welfare of Now and the Green (Post) Politics of the Future', *Critical Social Policy*, 31(2): 174–93. Published online 9 February 2011. Available at http://csp.sagepub.com/content/31/2/174.abstract. Accessed on 21 December 2011.

CCDPRJ (Civic Coalition for Defending Palestinian Rights in Jerusalem) (2009) *Aggressive Urbanism: Urban Planning and the Displacement of Palestinians within and from Occupied East Jerusalem*. Jerusalem: CCDPRJ.

Chalmers, T. (1821) *The Christian and Civic Economy of Large Towns, London 1856*. Glasgow: Chalmers and Collins.

China Daily Business Weekly, 'Some Basic Facts about the Three Gorges Dam'. Available at www.chinadam.com/dam/facts.htm. Accessed on 6 June 2011.

Christian Aid (2007) *Human Tide: The Real Migration Crisis*. London: Christian Aid.

Churchill, W. (1998) *A Little Matter of Genocide*. Winnipeg: Arbeiter Ring Publishing.

Cloward, R. and Piven, F. (1979) *Regulating the Poor: The Function of Public Welfare*. London: Tavistock.

Coates, J. (2003) *Ecology and Social Work: Towards a New Paradigm*. Halifax: Fernwood Books.

Coates, J. (2005) 'Environmental Crisis: Implications for Social Work', *Journal of Progressive Human Services*, 16(1): 25–49.

Coates, J., Gray, M. and Hetherington, T. (2006) 'An "Ecospiritual" Perspective: Finally, a Place for Indigenous Approaches', *British Journal of Social Work*, 36(3): 381–99.

Comerio, M. (2002) 'Housing Issues after Disasters', *Journal of Contingencies and Crisis Management*, 5(3): 166–73.

Connelly, S., Markey, S. and Roseland, M. (2011) 'Bridging Sustainability and the Social Economy: Achieving Community Transformation through Local Food Initiatives', *Critical Social Policy*, 31(2): 308–24.

Connett, M. (2003) *The Phosphate Fertilizer Industry: An Environmental Overview*. Available at www.fluoridealert.org/phosphate/overview.htm. Accessed 6 April 2011.

Conway, L. (2010) *A Case Study of The Isle of Eigg Heritage Trust, Scotland*. Available at www.caledonia.org.uk/socialland/eigg.htm. Accessed 2 August 2011.

Cox, A. (2000) 'Will Tribal Knowledge Survive the Millennium?' *Science 7, Essays on Science and Society*, 287(5450): 44–5.

Craig, G. and Mayo, M. (eds.) (1995) *Community Empowerment: A Reader in Participation and Development*. London: Zed Books.

Crawford, T. (2011) 'Demonstrators Plan to "Disrupt" Construction of Four-Lane Highway', *The Vancouver Sun*, 25 April, p. A10.

Cronin, M., Ryan, D. and Brier, D. (2007) 'Support for Staff Working in Disaster Situations: A Social Work Perspective', *International Social Work*, 50(3): 370–82.

Crouch, D. and Ward, C. (1997) *The Allotment: Its Landscape and Culture*. London: Five Leaves Publication.

Cryderman, K. (2009) 'Accord Reached on Global Warming,' *The Vancouver Sun*, 19 December, p. B1.

Culpitt, I. (1992) *Welfare and Citizenship: Beyond the Crisis of the Welfare State?* London: Sage.

Curtis, A., Li, B., Marx, B., Mills, J. and Pine, J. (2011) 'A Multiple Additive Regression Tree Analysis of Three Exposure Measures during Hurricane Katrina', *Disasters*, 35(1): 19–35.

Cyr, G. (2007) 'An Indigenist and Anti-Colonialist Framework for Practice' in L. Dominelli (ed.) *Revitalising Communities in a Globalising World*. Aldershot: Ashgate.

Dass-Brailsford, P. (2008) 'After the Storm: Recognition, Recovery and Reconstruction', *Professional Psychology: Research and Practice*, 39(1): 24–30.

De Moor, A. and Calamai, P. (1997) *Subsidizing Unsustainable Development: Undermining the Earth with Public Funds*. Ottawa: Institute for Research on Public Expenditure.

De Souza, M. (2011) 'Oilsands Pollution Major Concern', *The Vancouver Sun*, 26 December, p. A26.

Dempsey, J. and Ewing, J. (2011) 'Germany in Reverse, Will Close Nuclear Plants by 2022', *New York Times*, 30 May. Also available at http://www.nytimes.com/2011/05/31/world/europe/31germany.html. Accessed on 22 December 2011.

Denton, M. (1986) 'Environmentalism and Elitism: A Conceptual and Empirical Analysis', *Environmental Management*, 10(5): 581–9.

Delanty, G. (2003) *Community*. London: Routledge.

Dixon, J. (2011) 'Diverse Food Economies, Multivariant Capitalism, and the Community Dynamic Shaping Contemporary Food Systems', *Community Development Journal*, 46(1): 20–35.

Doh, J. P. and Teegen, H. (2003) *Globalization and NGOs: Transforming Business, Government and Society*. Westport, CT: Praeger Publishers.

Dominelli, L. (1997) *Sociology for Social Work*. London: Macmillan.

Dominelli, L. (2000) 'International Comparisons in Social Work' in R. Pearce and J. Weinstein (eds.) *Innovative Education and Training for Care Professionals: A Providers' Guide*. London: Jessica Kingsley.

Dominelli, L. (2002) *Anti-Oppressive Social Work Theory and Practice*. London: Palgrave-Macmillan.

Dominelli, L. (2004) *Social Work: Theory and Practice for a Changing Profession*. Cambridge: Polity Press.

Dominelli, L. (2006) *Women and Community Action*. Bristol: Policy Press.

Dominelli, L. (2010a) *Social Work in a Globalizing World*. Cambridge: Polity Press.

Dominelli, L. (2010b) 'Anti-Oppressive Practice' in M. Gray and S. Webb (eds.) *Ethics and Value Perspectives in Social Work*. London: Palgrave/Macmillan.

Dominelli, L. (2011) 'Climate Change: Social Workers' Roles and Contributions to Policy Debates and Interventions', *International Journal of Social Welfare*, 20(4/October): 430–8.

Dominelli, L. (2012). 'Social Work in Times of Disaster: Practising across Borders' in M. Kearnes, F. Klauser and S. Lane (eds) *Risk Research: Practice, Politics and Ethics*. Oxford: Wiley-Blackwell Publishers.

Dominelli, L. (in press) 'Mind the Gap: Built Infrastructures, Sustainable Caring Relations and Resilient Communities in Extrme Weather Events', *Australian Journal of Social Work*.

Dominelli, L., Strega, S., Warmsley, C., Callahan, M. and Brown, L. (2011) ' "Here's My Story": Fathers of Looked After Children Recount their Experiences of the Canadian Child Welfare System', *British Journal of Social Work*, 41(2): 351–67.

Donnelly, A. (2011) 'Say "No" to California Dreams, Nightmares', *The Vancouver Sun*, 29 April, p. A13.

Dovers, S. and Handmer, J. (1992) 'Uncertainty, Sustainability and Change', *Global Environmental Change*, 2(4): 262–76.

D'Silva, T. (2006) *The Black Box of Bhopal: A Closer Look at the World's Deadliest Industrial Disaster*. Victoria, BC: Trafford Publishing.

DuPuis, E., Goodman, D. and Harrison, J. (2006) 'Just Values or Just Value? Remaking the Local in Agro-food Studies' in T. Marsden and J. Murdoch (eds.) *Between the Local and the Global: Confronting Complexity in the Contemporary Agri-food Sector*. Amsterdam: Elsevier JAI.

EC (European Commission) (2004) *Joint Report on Social Inclusion*. Brussels: EC Directorate General for Employment and Social Affairs.

ECA (Economic Commission for Africa) (2005) *Review of the Application of Environmental Impact Assessment in Selected African Countries*. Addis Ababa: ECA.

Edwards, M., Hulme, D. and Wallace, T. (1999) *NGOs in a Global Future: Marrying Local Delivery to Worldwide Leverage*. Birmingham: Birmingham University Press.

Edwards, S. (2008) 'Social Breakdown in Darfur', *Forced Migration Review*, 31: 23–4.

Ehrenreich, J. (1985) *The Altruistic Imagination: A History of Social Work and Social Policy in the United States*. Ithaca, NY: Cornell University Press.

Emejulu, A. (2011) 'The Silencing of Radical Democracy in American Community Development: The Struggle of Identities, Discourses and Practices', *Community Development Journal*, 46(2): 229–44.

Engels, F. (1972) [1884] *The Origin of the Family, Private Property and the State*. London: Lawrence and Wishart Publishing.

Englund, W. (2011) '25 Years after Chernobyl, Nuclear Crisis, A Sense of Betrayal', *The Vancouver Sun*, 26 April, p. B4.

ERRC (European Roma Rights Centre) (2007) *Written Comments of the European Roma Rights Centre, the Centre on Housing Rights and Evictions, Osservazione and Sucar Drom Concerning Italy for Consideration by the United Nations Committee on the Elimination of Racial Discrimination at its 72nd Session*. Budapest: ERRC.

Escobar, A. (1998) 'Whose Knowledge, Whose Nature? Biodiversity, Conservation, and the Political Ecology of Social Movements', *Journal of Political Ecology*, 5: 53–82.

Escobedo, T. (2009) 'FBI Knew of Report that Suicide Bomber was Somali-American', *CNN*, 23 September.

Ewen, S. and Pusztai, A. (1999) 'Effects of Diets Containing Genetically Modified Potatoes Expressing *Galanthus nivalis* Lectin on Rat Small Intestine', *Lancet*, 354: 1353–4.

Fackler, M. (2011) 'Report Finds Japan Underestimated Tsunami Danger', *New York Times*, 1 June.

Falstrom, D. (2001) 'Stemming the Flow of Environmental Displacement: Creating a Convention to Protect Persons and Preserve the Environment', *Journal of International Environmental Law and Policy*, 1: 1–19.

Figley, C. R. (eds.) (1995) *Compassion Fatigue: Coping with Secondary Traumatic Stress Disorder in Those who Treat the Traumatized*. New York: Brunner/Mazel.

Fisher, S. (2005) *Gender-Based Violence in Sri Lanka in the Aftermath of the Tsunami Crisis*. Leeds: University of Leeds.

Flexner, A. (2001) [1915] 'Is Social Work a Profession?' Reprinted in *Research on Social Work Practice*, 11(2): 152–65.

Florida, R. (2004) *The Rise of the Creative Class: And How It Is Transforming Work, Leisure, Community and Everday Life*. New York: Basic Books.

FNEATWG (First Nations Environmental Assessment Technical Working Group) (undated) *First Nations Environmental Assessment Toolkit*. Cranbrook, BC: Canadian Columbia River Inter-tribal Fisheries Commission. Also available at www.fneatwg.org/toolkit.html. Accessed 12 August 2011.

Folghereiter, F. (2003) *Relational Social Work: Toward Network and Societal Practices*. London: Jessica Kingsley.

Folke, C., Carpenter, T., Elmqvist, L., Gunderson, C., Holling, B., Walker, J., Bengtsson, F., Berkes, J., Colding, K., Danell, M., Falkenmark, L., Gordon, R., Kaspersson, N., Kautsky, A., Kinzig, S., Levin, K. G., Mäler, F., Moberg, L., Ohlsson P., Olsson, E., Ostrom, W., Reid, J., Rockstroem, H., Savenije, U. and Svedin, U. (2002) *Resilience and Sustainable Development: Building Adaptive Capacity in a World of Transformations*. Stockholm: Environmental Advisory Council to the Swedish Government.

Forbes, T. (2011) 'Kachin NGO Questions Myitsone Dam Suspension', *Healthy Rivers, Happy Communities for Now and the Future*, 17 October. Available at www.burmariversnetwork.org/news/11-news/723-kachin-ngo-questions-myitsone-dam-suspension.html. Accessed 23 October 2011.

Foster, P. (2010) 'Deadly Landslide was Avoidable, Experts Say', *The Vancouver Sun*, 12 August, p. B4.

Franklin, J. (2011) 'For Chile's Rescued Miners, a New Darkness', *The Globe and Mail*, 5 August, p. A12.

Friends of the Earth (2008) *A Dangerous Distraction: Why Offsetting is Failing the Earth and People: The Evidence*. London: Friends of the Earth.

Garrett, P. M. (2009) 'Recognizing the Limitations of the Political Theory of Recognition: Axel Honneth, Nancy Fraser and Social Work', *British Journal of Social Work*, 40: 1517–33.

Germain, C. (ed.) (1979) *Social Work Practice: People and Environments, An Ecological Perspective*. New York: Columbia University Press.

Germain, C. and Gitterman, A. (1995) *The Life Model of Social Work Practice: Advances in Theory and Practice*. New York: Columbia University Press.

GHA (Global Humanitarian Assistance) (2010) *Global Humanitarian Assistance Report, 2010*. Wells, Somerset, UK: Development Initiative.

Giddens, A. (2009) *The Politics of Climate Change*. Cambridge: Polity.

Gilchrist, R. and Jeffs, T. (2001) *Settlements, Social Change and Community Action: Good Neighbours*. London: Jessica Kingsley Publishers Ltd.

Gill, O. and Jack, G. (2007) *Child and Family in Context: Developing Ecological Practice in Disadvantaged Communities*. Lyme Regis: Russell House.

Glover, S. (2011) 'Forests Require Flexible Management', *The Vancouver Sun*, 26 April, p. A9.

Golders, J. (2011) 'Obama Cooling on Global Warming', *The Vancouver Sun*, 29 April, p. A13.

Goodman, D. and DuPuis, E. M. (2002) 'Knowing Food and Growing Food: Beyond the Production–Consumption Debate in the Sociology of Agriculture'. *Sociologia Ruralis*, 42(1): 6–23.

Gottschalk, J. (ed.) (1993) *Crisis Response: Inside Stories on Managing under Siege*. Detroit: Visible Ink Press.

Grande, S. (2004) *Red Pedagogy: Native American Political Thought*. Lanham, MD: Rowman & Littlefield.

Gray, M., Coates, J. and Yellowbird, M. (eds.) (2008) *Indigenous Social Work Around the World: Towards Culturally Relevant Education and Practice*. Aldershot: Ashgate.

Green, J. and Thomas, R. (2007) 'Learning Through Our Children, Healing For Our Children: Best Practice in First Nations Communities' in L. Dominelli (ed.) *Revitalising Communities in a Globalising World*. Aldershot: Ashgate.

Greenough, P. (2008) 'Burden of Disease and Health Status among Hurricane Katrina Displaced Persons in Shelters: A Population-based Cluster Sample', *Annals of Emergency Medicine*, 51(4): 426–32.

Gubbins, N. (2010) *The Role of Community Energy Schemes in Supporting Community Resilience*. York: Joseph Rowntree Foundation.

Guzmán, J., Martine, G., McGranahan, G., Schensul, D. and Tacoli, C. (eds.) (2009) *Population Dynamics and Climate Change*. New York: UNFPA/IIED.

Haddad, L. and Godfray, C. (2011) *Global Food and Farming Futures Report*. Oxford: Institute of Development Studies.

Hagerty, J. (2011) 'Shale-Gas Boom Fuels Race for Factories', *The Globe and Mail*, 27 December, p. B6.

Haig-Brown, C. (1988) *Resistance and Renewal: Surviving the Indian Residential School*. Vancouver: Arsenal Pulp Press.

Hallegatte, S., Henriet, F., Patwardhan, A., Narayanan, K., Ghosh, S., Karmakar, S., Patnaik, U., Abhagankar, A., Pohit, S., Corfee-Morlot, J. and Herweiger, C. (2010) *Flood Risks, Climate Change Impacts and Adaptation Benefits in Mumbai: An Initial Assessment of Socio-Economic Consequences of Present and Climate Change Induced Risks and of Possible Adaptation*. OECD Environment Working Paper 27. Paris: OECD Publishing.

Hanson, L. (2009) 'Environmental Justice as a Politics of Place: An Analysis of Five Canadian Environmental Groups' Approaches to Agro-Food Issues' in J. Agyeman, P. Cole, R. Haluza-Delay and P. O'Riley (eds.) (2009) *Speaking for Ourselves: Environmental Justice in Canada*. Vancouver: University of British Columbia Press.

Hoang, X. T., Dang, A. N. and Tacoli, C. (2005) *Livelihood Diversification and Rural–Urban Linkages in Vietnam's Red River Delta*. Rural–Urban Working Paper 11. London: IIED.

Hodgson, D. (2002) 'Introduction: Comparative Perspectives on Indigenous Rights Movements in Africa and the Americas', *American Anthropologist*, 104(4): 1037–49.

Hoff, M. and McNutt, G. (1994) *The Global Environmental Crisis: Implications for Social Welfare and Social Work*. Brookfield, VT.: Avebury.

Hoogvelt, A. (2007) 'Globalisation and Imperialism: Wars and Humanitarian Intervention', in L. Dominelli (ed.) *Revitalising Communities in a Globalising World*. Aldershot: Ashgate.

Houston, S. (2008). 'Beyond Homo Economicus: Recognition, Self-realization and Social Work', *British Journal of Social Work*, 40(3): 841–57.

Howard, E. (1902) *Garden Cities of Tomorrow*. London: S. Sonnenschein and Co., Ltd.

Hudson, C. (2000) 'At the Edge of Chaos: A New Paradigm for Social Work', *Journal of Social Work Education*, 36(2): 215–30.

Humber, Y. and Kate, D. T. (2011) 'Mongolia Lays Tracks for the Future', *The Vancouver Sun*, 25 April, p. B5.

Hyde, M., Dixon, J. and Drover, G. (2003) 'Welfare Retrenchment or Collective Responsibility? The Privatisation of Pension Provisions in Western Europe', *Social Policy and Society*, 2(3): 189–97.

IAIA (International Association for Impact Assessment) (2010) *Social Impact Assessment*. Available at www.iaia.org/iaiawiki/sia.ashx. Accessed 11 August 2011.

ICG (International Crisis Group) (2002) *Border Disputes and Conflict Potential: ICG Asia Report Number 33*. Brussels: ICG.

Ioakimides, V. (2010) 'Expanding Imperialism, Exporting Expertise: International Social Work and the Greek Project', *International Social Work*, 54(4): 505–19.

IOM (International Organisation for Migration) (2010) *The Future of Migration: Building Capacity for Change*. London: IOM.

IPCC (Intergovernmental Panel on Climate Change) (2001) *Glossary of Terms Used by Working Group II – Impacts, Adaptation, and Vulnerability in the Third Assessment Report*. Available at www.grida.no/climate/ipcc_tar/wg2/689.htm. Accessed on 2 April 2011.

IPCC (2007) *Fourth Assessment Report of the Intergovernmental Panel on Climate Change*. Geneva: IPCC.

ISDR (2005) *The Hyogo Framework for Action 2005–2015: Building the Resilience of Nations and Communities to Disasters*. New York: United Nations, ISDR.

Jackson, B. (1993) 'Union Carbide: Disaster at Bhopal' in J. Gottschalk (ed.) *Crisis Response: Inside Stories on Managing under Siege*. Detroit: Visible Ink Press.

Jeong, M. (2011) 'In Mumbai, Dhobis against Developers', *The Globe and Mail*, 4 August, p. B6.

Jiang, L. and Hardee, K. (2009) *How Do Recent Population Trends Matter to Climate Change?* Working Paper 1. Washington, DC: Population Action International.

Jones, K. and Duarte-Davidson, R. (1997) 'Transfers of Airborne PCDD/Fs to Bulk Deposition Collectors and Herbage', *Environmental Science and Technology*, 31(10): 2937–43.

Jones, V. (2008) *The Green Collar Economy: How One Solution Can Fix Our Two Biggest Problems*. New York: Harper Collins.

Kameri-Mbote, P. (2007) 'Water, Conflict and Cooperation: Lessons from the Nile River Basin', *Navigating Peace* (Woodrow Wilson International Center for Scholar, Washington, DC), 4: 1–6. Also available at www.wilsoncenter.org/topics/pubs/NavigatingPeaceIssue4.pdf. Accessed 28 December 2011.

Kan, N. (2011) 'The Road to Recovery and Rebirth', *The Vancouver Sun*, 27 April, p. A15.

Kaseke, E. (1996) 'Social Work and Social Development in Zimbabwe', *Journal of Social Development in Africa*, 11(2): 151–65.

Kendall, K. (2000) *Social Work Education: Its Origins in Europe*. Alexandria, VA.: Council on Social Work Education.

Khan, U. (2008) 'The Number of Workers Spending More Than One Hour Commuting to Work Falls', *The Daily Telegraph*, 27 October, p. 32. Also available at www.telegraph.co.uk/motoring/3269119/The-number-of-workers-spending-more-than-an-hour-commuting-to-work-falls.html. Accessed on 30 December 2011.

Kirkpatrick, D. and Bryan, M. (2007) 'Hurricane Emergency Planning by Health Care Providers Serving the Poor', *Journal of Health Care*, 18(2): 299–314.

Klein, N. (2008) *The Shock Doctrine: The Rise of Disaster Capitalism*. London: Allen Lane.

Klein, R., Nicholls, R. and Thomalla, F. (2004) *Resilience to Natural Hazards: How Useful is the Concept?* EVA Working Paper 9. Potsdam: Potsdam Institute for Climate Change Research.

Klinenberg, E. (2002) *Heat Wave: A Social Autopsy of Disaster in Chicago*. Chicago: University of Chicago Press.

Kovats, R. S. and Ebi, K. L. (2006) 'Heatwaves and Public Health in Europe', *European Journal of Public Health*, 16(6): 522–99.

Kroll, L. and Fass, A. (2011) *Forbes Magazine*, Special Report, *The World's Billionaires*, 8 March.

Kuecker, G., Mulligan, M. and Nadarajah, Y. (2011) 'Turning to Community in Times of Crisis: Globally Derived Insights on Local Community Formation', *Community Development Journal*, 46(2): 245–64.

Laming, H. (2009) *The Protection of Children in England: A Progress Report*. London: DCSF.

Landon, J. (2011) 'Greek Bonds Lure Some, Despite Risk', *New York Times, Global Business*, 28 September. Available at www.nytimes.com/2011/09/29/business/global/hedge-funds-betting-on-lowly-greek-bonds.html?pagewanted=all. Accessed 10 October 2011.

Lane, S. (2008) 'Climate Change and the Summer 2007 Floods in the UK', *Geography*, 93(2): 91–7.

Lane, S., Odoni, N., Landström, C., Wahtmore, S. J., Ward, N. and Bradley, S. (2011) 'Doing Flood Risk Science Differently: An Experiment in Radical Scientific Method', *Transactions of the Institute of British Geographers*, 36(1): 15–36.

Larsen, J. (2003) *Record Heat Wave in Europe Takes 35,000 Lives: Far Greater Loses May Lie Ahead*. Washington, DC: Earth Policy Institute.

Leckie, S. (2009) 'Climate-Related Disasters and Displacement: Homes for Lost Homes:Lands for Lost Lands' in J. Guzmán, G. Martine, G. McGranahan, D. Schensul and C. Tacoli (eds.) *Population Dynamics and Climate Change*. New York: UNFPA/IIED.

Leggett, M. (2011) 'Dangers of Fracking Go Beyond Poisoned Water and Earthquakes', *Earth Times*, 22 March. Available at www.earthtimes.org/energy/dangers-hydraulic-fracturing-poisoned-water-supplies-earthquakes/552/. Accessed 12 December 2011.

Levesque, L. and Mathieu, A. (2011) 'Tunnel Collapse Prompts Montreal Mayor to Call for Immediate Funds', *The Globe and Mail*, 3 August, p. A9.

Levy, D. (2010) *IMF Backtracks on Debt Relief for Haiti*. Available at www. naomiklein.org/articles/2010/01/imf-backtracks-debt-relief-haiti. Accessed 5 June 2011.

Liebenthal, A. (2005) *Extractive Industries and Sustainable Development*. Washington, DC: World Bank.

Lorenz, W. (1994) *Social Work in a Changing Europe*. London: Routledge.

Löscher, P. (2009) *Pictures of the Future: Magazine for Research and Innovation: Special Edition: Green Technologies*. Munich: Siemens, Aktiengesellschaft.

Macalister, T. and Webb, T. (2011) 'Carbon Fraud May Force Longer Closure of EU Emissions Scheme', *The Observer*, 23 January, p. 20.

Macqueen, A. (2011) 'Los 33: Chilean Miners Face Up to a Strange New World', *The Observer*, 17 July. Available at www.guardian.co.uk/world/2011/jul/17/ chilean-miners-one-year-on. Accessed 5 August 2011.

Malthus, T. R. (1798) *An Essay on Principles of Population*. Oxford World Classics Reprint. Oxford: Oxford University Press.

Manthorpe, J. (2010) 'Floods Threaten Fabric of the Pakistani State', *The Vancouver Sun*, 11 August, p. B1.

Manyena, S. B. (2006) 'The Concept of Resilience Revisited', *Disasters*, 30(4): 434–50.

Maracle, L. (1993) *Ravensong*. Vancouver: Press Gang Publishers.

Martell, P. (2011) 'Camps Fill as Somalia's Famine Spreads', *The Globe and Mail*, 4 August, p. A16.

Marx, K. (1978) *Das Capital: A Critique of Political Economy*. Moscow: Progress Publishers.

Mary, N. L. (2008) *Social Work in a Sustainable World*. New York: Lyceum Books.

Mason, R. (2009). 'UN Pleads for Investment Deals at Copenhagen,' *The Daily Telegraph*, 9 December, p. 83.

Massey, D., Axinn, W. and Ghimire, D. (2007) *Environmental Change and Out-Migration: Evidence from Nepal*. Population Studies Center Research Report 07-615. Ann Arbor, MI: Institute for Social Research, University of Michigan.

McCurry, J. (2011) 'Japan Faces Power Shortages Due to Nuclear Shutdowns', *The Guardian*, 6 July. Available at www.guardian.co.uk/world/2011/jul/06/ japan-power-shortages-nuclear-shutdowns. Accessed on 12 December 2011.

McInerney-Lankford, S., Darrow, M. and Rajamani, L. (2011) *Human Rights and Climate Change: A Review of the International Legal Dimensions*. World Bank Study. Washington, DC: World Bank.

McKinnon, J. (2008) 'Exploring the Nexus Between Social Work and the Environment', *Australian Social Work*, 61(3): 268–82.

McMichael, A. J., Woodruff, R. and Hales, S. (2006) 'Climate Change and Human Health: Present and Future Risks', *The Lancet*, 367: 859–69.

Medvedev, G. (1991) [1989] *The Truth about Chernobyl*. Trans. E. Rossiter. New York: Basic Books.

Miller, S. M., Rein, M. and Levitt, P. (1990) 'Community Action in the United States', *Community Development Journal*, 25(4): 365–8.

Mitchell, T., Tanner, T. and Haynes, K. (2009) *Children as Agents of Change for Disaster Risk Reduction: Lessons from El Salvador and the Philippines.* Brighton: IDS.

Mohanty, C. (2003) *Feminism without Borders: Decolonizing Theory, Practising Solidarity.* Durham, AL: Duke University Press.

Mott, A. (2004) 'Increasing Space and Influence Through Community Organising and Citizen Monitoring: Experiences from the USA', *Institute of Development Studies Bulletin,* Special Issue, *New Democratic Spaces,* 35(2): 91–112.

Munro, E. (2011) *Munro Review of Child Protection: Final Report – A Child-Centred System.* London: Department of Education.

Murphy, M. (2006) *Sick Building Syndrome and the Problems of Uncertainty: Environmental Politics, Technoscience and Women Workers.* Durham, AL: Duke University Press.

Myer, N. (2005) 'Environmental Refugees: An Emerging Security Issue'. Paper presented at the 13th Economic Forum, Organization for Security and Co-operation in Europe, Prague, 23–27 May.

Nagle, L. E. (2008) 'Selling Souls: The Effects of Globalization on Human Trafficking and Forced Servitude', *Wisconsin International Law Journal,* 26(1): 131–40.

Närhi, K. (2004) 'The Eco-social Approach in Social Work and the Challenges to the Expertise of Social Work'. Ph.D. thesis, University of Jyväskylä, Jyväskylä, Finland, 30 June.

Newbery, D. (1997) 'Determining the Regulatory Asset Base for Utility Price Regulation', *Utilities,* 6(1): 1–8. Nicholls, R., Hanson, S., Herweijer, C., et al. (2007) *Ranking of the World's Cities Most Exposed to Coastal Flooding Today and in the Future.* Southampton: Southampton University and OECD.

Nicholls, R., Hanson, S., Herweijer, C., Patmore, N., Hallegatte, S., Corfee-Morlot, J., Chateau, J. and Muir-Wood, R. (2007) *Ranking of the World's Cities Most Exposed to Coastal Flooding Today and in the Future.* Southampton: Southampton University and OECD.

Nickel, S. (2011) SaskPower to build first carbon capture plant, *The Vancouver Sun,* 27 April, p. C11.

OCSE (Office for Security and Co-operation in Europe) (2009) *Assessment of the Human Rights of Roma and Sinti.* Warsaw and The Hague: OCSE and Office for Democratic Institutions and Human Rights, High Commissioner for National Minorities.

ONS (Office for National Statistics) (2009) *Ageing in the UK: Interactive Mapping Tool.* Available at www.statistics.gov.uk/ageingintheuk/default.htm. Accessed 2 July 2011.

ONS (2011) *Population Estimates.* Available at www.statistics.gov.uk/cci/nugget. asp?id=6. Accessed 7 August 2011.

Osif, B., Baratta, A. and Conkling, T. W. (2004) *TMI 25 Years Later: The Three Mile Island Nuclear Power Plant Accident and Its Impact.* University Park, PA: Pennsylvania State University Press.

Owen, D. (1982) *The Government of Victorian London, 1855–1889: The Metropolitan Board of Works, the Vestries and the City Corporation.* Cambridge, MA: Belknap Press.

Oxtoby, K. (2009) 'Social Pedagogues in Mainland Europe', *Community Care*, 18 March, p. 21.

Padilla, B. (2008) *In Philippines Slums Desperately Poor Sell Kidneys for Cash*. Available at http://afp.google.com/article/ALeqM5hArQn4tsN4n_cT1TcmiLJN_axsfQ. Accessed on 2 August 2011.

Panet-Raymond, J. (1991) *Partnership or Paternalism?* Montreal University of Montreal Publication.

Paperny, A. (2011) 'Companies Vie for Iconic Reactor', *The Globe and Mail*, 4 August, p. A4.

Pardeck, J. (1996) *Social Work Practice: An Ecological Approach*. Westport, CT: Auburn House.

Parry, N., Rustin, M. and Satyamurti, C. (eds.) (1979) *Social Work, Social Welfare and the State*. London: Edward Arnold.

Parry, R. (1989) *Privatisation: Social Work Research Highlights 18*. London: Jessica Kingsley Publishers.

Pearsall, H. and Pierce, J. (2010) 'Urban Sustainability and Environmental Justice: Evaluating the Linkages in Public Planning/Policy Discourse', *Local Environment*, 15(6): 569–80.

Penner, D. (2010) 'Kootenay Partners Forge Ahead with $900-million Power Project', *The Vancouver Sun*, 27 August, p. C1.

Peskett, L., Huberman, D., Bowen-Jones, E., Edwards, G. and Brown, J. (2008) *Making REDD Work for the Poor: Poverty Environment Partnership (PEP) Policy Brief*. Brighton: ODI.

Pickles, E. (2011) *Action to Boost Support for Voluntary Sector and Cut Red Tape for Councils*. Available at www.communities.gov.uk/news/corporate/1885482. Accessed on 26 April 2011.

Pisupati, B. (2004) *Connecting the Dots: Biodiversity, Adaptation, Food Security and Livelihoods*. Nairobi: BLL, DELC, UNEP.

Pitt, M. (2007) *The Pitt Review: Lessons Learnt from the 2007 Floods*. London: Cabinet Office and Environment Agency.

Plummer, J. (2011) 'Analysis: Sector Budgets Feel the Squeeze', *Third Sector*. Available at www.thirdsector.co.uk/news/Article/1060843/Analysis-Sector-budgets-feel-squeeze/. Accessed on 26 April 2011.

Ponce de Leon, M. (2008) 'Social Work in Chile'. Paper given at the IASSW Board Seminar, International Conference Centre, Durban, South Africa, 10–13 July.

Putnam, R. (2000) *Bowling Alone: The Collapse and Revival of American Community*. New York: Simon and Schuster.

Pyles, L. (2007) 'Community Organising for Post-Disaster Development: Locating Social Work', *International Social Work*, Special Issue on Disasters, 53(5): 321–33.

Pynn, L. (2011) 'Fisheries' "Evasive" Action Costs $80,000 in Orca Case', *The Vancouver Sun*, 27 April, p. A3.

Quayle, M. and Richards, J. (2011) 'BC Needs to Take Green Economy to Next Level', *The Vancouver Sun*, 26 April, p. A9.

Ramon, S. (ed.) (2008) *Social Work in the Context of Political Conflict*. Birmingham: Venture Press.

Ramon, S., Campbell, J., Lindsay, J., McCrystal, P. and Baidoun, N. (2006) 'The Impact of Political Conflict on Social Work: Experiences from Northern Ireland, Israel and Palestine', *British Journal of Social Work*, 36: 435–50.

Ravallion, M., Chen, S. and Sangraula, P. (2008) *Dollar a Day Revisited*. Policy Research Working Paper 4620. Washington, DC: World Bank.

Reacher, M., McKenzie, K., Lane, C., Nichols, T., Iversen, A., Hepple, P., Walter, T., Laxton, C. and Simpson, J. (2004) 'Health Impacts of Flooding in Lewes: A Comparison of Reported Gastrointestinal and Other Illness and Mental Health in Flooded and Non-flooded Households', *Communicable Disease and Public Health*, 7(1): 1–8.

Roberts, G. (1998) 'Environmental Justice and Community Empowerment: Learning from the Civil Rights Movement', *American University Law Review*, 48: 229–57.

Rogge, M. E. (1994). 'Environmental Justice: Social Welfare and Toxic Waste' in M. D. Hoff and J. G. McNutt (eds.) *The Global Environmental Crisis Implications for Social Welfare and Social Work*. Aldershot: Ashgate.

Rogge, M. E. and Coombs-Orme, T. (2003). 'Protecting Children from Chemical Exposure: Social Work and US Social Welfare Policy'. *Social Work*, 48(4): 439–50.

Rogge, M. E. and Darkwa, O. K. (1996). 'Poverty and the Environment: An International Perspective for Social Work', *International Social Work*, 39: 395–409.

Room, G. (1995) *Beyond the Threshold: The Measurement and Analysis of Social Exclusion*. Bristol: Policy Press.

Rose, F. (2000) *Coalitions across Class Divides*. Ithaca, NY.: Cornell University Press.

Roseland, M. (2005) *Toward Sustainable Communities: Resources for Citizens and their Governments*. Gabriola Island, BC: New Society Publishers.

Rosen, A. and Livne, S. (2011) *Personal versus Environmental Emphases in Social Workers' Perceptions of Client Problems*. Available online at www.mcnellie.com/525/readings/rosenlivne.pdf. Accessed on 26 April 2011.

Rosenhek, R. (2006) 'Earth, Spirit and Action: The Deep Ecology Movement as Spiritual Engagement', *The Trumpeter: The Journal of Ecosophy*, 22(2): 90–5.

Rounsevell, M. and Reay, D. (2009) 'Land Use and Climate Change in the UK', *Land Use Policy Journal*, 265: 5160–9.

Roy, A. (1999) *The Greater Common Good*. Available at www.narmada.org/gcg/gcg.html. Accessed 6 June 2011.

Ruwhiu, L. (1998) '*Te Puawaitango o te ihi me te wehi*': The Politics of Maori Social Policy Development. Ph.D. thesis, Massey University, Palmerston North.

Schlosberg, D. (2007) *Environmental Justice: Theories, Movements and Nature*. Oxford: Oxford University Press.

Schlosser, E. (2001) *Fast Food Nation: The Dark Side of the All American Meal*. New York Houghton Mifflin Co.

Schmidt, G., Westhhuies, A., Lafrance, J. and Knowles, A. (2001) 'Social Work in Canada: Results from a National Sector Study', *Canadian Social Work*, 3(2): 83–92.

Schumacher, E. F. (1974) *Small is Beautiful: A Study Of Economics As If People Mattered*. London: ABACUS.

Schumpeter, J. (1935) 'The Analysis of Economic Change', *Review of Economic Statistics* (May): 2–10.

Schutz, A. and Sandy, M. (2011) *Collective Organising for Social Change: An Introduction to Community Organising*. London: Palgrave.

SCIE (Social Care Institute for Excellence) (2008) *Research: Role of Social Work in Mental Health Services*. Available at www.communitycare.co.uk/Articles/14/07/2008/108835/Research-role-of-social-work-in-mental-health-services.htm. Accessed on 20 October 2011.

Scoones, I. (1999) 'New Ecology and the Social Sciences: What Prospects for a Fruitful Engagement?' *Annual Review of Anthropology*, 28: 479–507.

Scudder, T. (2005) *The Kariba Case Study*. Social Science Working Paper 1227. Pasedena, CA: California Institute of Technology.

Seabrook, J. (2007) *Cities: Small Guides to Big Issues*. London: Pluto Press.

Sebellos, F., Tanner, T., Tarazona, M. and Gallegos, J. (2011) *Children and Disasters: Understanding Impact and Enabling Agency*. Paris: UNICEF.

Sewpaul, V. and Hölscher, D. (2005) *Social Work and Neo-Liberalism*. Pretoria: Van Shaik.

Shah, A. (2010) *Poverty: Facts and Stats*. Available at www.globalissues.org/article/26/poverty-fact-and-stats. Accessed on 4 July 2011.

Sharkey, P. (2007) 'Survival and Death in New Orleans: An Empirical Look at the Human Impact of Katrina', *Journal of Black Studies*, 37(4): 482–501.

Shaw, M. (2011) 'Stuck in the Middle? Community Development, Community Engagement and the Dangerous Business of Learning Our Democracy', *Community Development Journal*, 46(2): 128–46.

Shiva, V. (2003) 'Food Rights, Free Trade and Fascism' in M. Gibney (ed.) *Globalizing Rights*. Oxford: Oxford University Press.

Shivii, I. (2006) *The Silences Behind the NGO Discourse: The Role and Future of NGOs in Africa*. Oxford: Fahamu Ltd.

Shragge, E. and Fontane, J. M. (eds.) (2000) *Social Economy: International Debates and Perspectives*. Montreal: Black Rose Books.

Sinesi, B. and Ulph, D. (1998) *Species Loss through the Genetic Modification of Crops*. London: UCL.

Skehill, C. (2008) 'Looking Back While Moving Forward: The History of Social Work: Historical Perspectives in Social Work', *BJSW*, Special Issue Editorial, 38(4).

Spickett, J., Brown, H. and Katscherian, D. (2008) *Health Impacts of Climate Change: Adaptation Strategies for Western Australia*. Available at www.public.health.wa.gov.au/cproot/1510/2/Health_Impacts_of_Climate_Change.pdf. Accessed on 6 June 2011.

Steadman Jones, G. (1984) *Outcast London*. London: Pantheon Books.

Stern, D. (2010) 'Kyrgyzstan: Surviving Ethnic Conflict', *Global Post*, 17 June. Available at www.globalpost.com/dispatch/asia/100617/kyrgyzstan-news-ethnic%20conflict. Accessed on 4 August 2011.

Stern, N. (2006) *Stern Review of the Economics of Climate Change*. Cambridge: Cambridge University Press.

Stern, P., Bietz, T. and Guagnano, G. (1995) 'The New Ecological Paradigm in Social-Psychological Context', *Environment and Behaviour*, 27(6): 723–43.

Stier, M. (2011) *Rebuilding our Neighbourhoods Block by Block*. Available at www.stier.net/writing/other/Rebuilding_Our_Neighborhoods_Block_by_Block.pdf. Accessed on 28 April 2011.

Stott, K. (2009) 'Remote Village Turns to the Sun for Power',*Vancouver Sun*, 26 October, p. B4.

Sturge, S. (2010) 'Social Work in Palestine'. Paper presented to BASW group on Palestine.

Swift, K. and Callahan, M. (2010) *At Risk: Social Justice in Child Welfare and Other Human Services*. Toronto: Toronto University Press.

Tacoli, C. (2009) 'Crisis or Adaptation? Migration and Climate Change in a Context of High Mobility' in J. Guzmán, G. Martine, G. McGranahan, D. Schensul and C. Tacoli (eds.) *Population Dynamics and Climate Change*. New York: UNFPA/IIED.

Tait-Rolleston, W. and Pehi-Barlow, S. (2001) 'A Maori Social Work Construct' in L. Dominelli, W. Lorenz and H. Soydan (eds.) *Beyond Racial Divides: Ethnicities in Social Work*. Aldershot: Ashgate.

Tholfsen, T. (1976) *Working-Class Radicalism in mid-Victorian England*. London: Croom Helm Ltd.

Thompson, E. P. (1963) *The Making of the English Working Class*. London: Penguin.

Toomey, A. (2011) 'Empowerment and Disempowerment in Community Development Practice: Eight Roles Practitioners Play', *Community Development Journal*, 46(2): 181–95.

Townsend, M. (2011) 'A Stone's Throw from Dale Farm, Travellers' Stand-off Simmers On', *The Observer*, 30 October, p. 21.

TRRT (Thames Rivers Restoration Trust) (2011) *Projects*. Available at www.trrt.org.uk/index.aspx?articleid=15819. Accessed on 2 August 2011.

TW (Thames Water) (2011) *London's Victorian Sewer System*. Available at www.thameswater.co.uk/cps/rde/xchg/prod/hs.xsl/10092.htm. Accessed on 12 December 2011.

TWN (Third World Network) (2010) *Bonn Climate News Updates, May/June 2010*. Penang, Malaysia: Jutaprint.

UN (2003) *The Challenge of Slums*. New York: Human Development Programme.

UN (2006) *On Better Terms: A Glance at Key Climate Change and Disaster Risk Reduction Concepts*. New York: UN.

UN (2011) *The World Urbanisation Prospects: Report for 2009*. New York: UN Department of Economic and Social Affairs.

UNDP (United Nations Development Fund) (1998) *The Human Development Report, 1997*. New York: UNDP.

UNDP (United Nations Development Fund) (2008) *The Human Development Report, 2007*. New York: UNDP.

UNDP (United Nations Development Fund) (2009) *The Human Development Report, 2008*. New York: UNDP.

UNDP (United Nations Development Fund) (2011) *The Human Development Report, 2010*. New York: UNDP.

UNEP (United Nations Environment Programme) (2009) *Disasters and Conflicts: Factsheets*. New York: UNEP.

UNESA (United Nations Department of Economic and Social Affairs) (2009) *World Population Prospects: The 2008 Revision*. New York: UNESA.

UNESCO (United Nations Educational, Scientific and Cultural Organisation) (2006) *Water Conflicts: An Analysis of Water-Related Unrest and Conflict in Urban Context*. Paris: UNESCO/IHP.

Ungar, M. (2002) 'A Deeper, More Social Ecological Social Work Practice', *Social Service Review*, 76(3): 480–97.

UNICEF (United Nations International Children's Emergency Fund) (2010) *Recovery for All: A Call for Collective Action*. New York: UNICEF.

UNISDR (United Nations International Strategy for Disaster Reduction) (2004) *Living with Risk: A Global Review of Disaster Reduction Initiatives*. New York: UN.

Unruh, M. K. (2005) *Environmental Change and its Implication for Population Migration*. Cambridge: Cambridge University Press.

Van Wormer, K., Besthorn, F. and Keefe, T. (2007) *Human Behavior and the Social Environment: Macro-level Groups, Communities and Organizations*. New York: Oxford University Press. Second edition, 2010.

Velásquez, L. (2005) 'The Bioplan: Decreasing Poverty in Manizales, Colombia, through Shared Environmental Management' in S. Bass (ed.) *Reducing Poverty and Sustaining the Environment: The Politics of Local Engagement*. London: Earthscan.

Vergara, E. (2011) 'Protesters Throw Fruit at Chile's Rescued Miners', *Associated Press*, 6 August. Available on www.washingtonpost.com/world/protesters-throw-fruit-at-chiles-rescued-miners/2011/08/06/gIQAunxUzI_story.html. Accessed on 6 August 2011.

Vidal, J. (2010) 'Climate Aid Threat to Countries that Refuse to Back Copenhagen Accord', *The Observer*, 11 April.

Walker, A. (1990) 'The "Economic" Burden of Ageing and the Prospect of Intergenerational Conflict', *Ageing and Society*, 10: 377–96.

Walker, J. S. (2004) *Three Mile Island: A Nuclear Crisis in Historical Perspective*. Berkeley: University of California Press.

Walker, P. (2011) *Getting Humanitarian Aid Right*. Available at http://sites.tufts.edu/gettinghumanitarianaidright/. Accessed on 26 November 2011.

Walker, R. (2010) 'Global Population Still a Problem', *The Guardian*, 13 July.

Walter, M. (2008) 'Chavez's Cheap Oil Gives Him Sway Over US Allies, Aid Funds', *Bloomberg*, 5 October. Available at www.bloomberg.com/apps/news?pid=newsarchive&refer=Latin_America&sid=aUMDzYgZ0h6Y. Accessed on 12 June 2010.

Walton, R. (1975) *Women in Social Work*. London: Routledge and Kegan Paul.

Warmington, P., Daniels, H., Edwards, A., Leadbetter, J., Martin, D., Brown, S., and Middleton, D. (2004) 'Learning In and For Interagency Working: Conceptual Tensions in "Joined Up" Practice'. Paper presented to the TLRP Annual Conference, Cardiff, November.

Warren, M. (2001) *Dry Bones Rattling: Community Building to Revitalize American Democracy*. Princeton, NJ: Princeton University Press.

Waterhouse, R. (1994) 'Supervision for Safari Boy', *The Independent*, 13 January, and '"Safari Boy" Jailed for Nine Months', *The Independent*, 24 September.

Watts, J. (2011a) 'China Warns of Urgent Problems Facing Three Gorges Dam', *The Guardian*, 20 May, p. 24.

Watts, J. (2011b) 'Aung San Suu Kyi: China's Dam Project in Burma is "Dangerous and Divisive"', *The Guardian*, 12 August, p. 26.

Webb, B. (1909) *Minority Report*. London: Commission on the Poor Law.

Webb, S. (1918) *Labour and the New Social Order*. London: The Fabian Society.

Webb, S. (2010) '(Re)assembling the Left: The Politics of Redistribution and Recognition in Social Work', *British Journal of Social Work*, 40(8): 2364–79.

Webster, M., Ginnetti, K., Walker, P., Coppard, D. and Kent, R. (2009) 'The Humanitarian Response Costs of Climate Change', *Journal of Environmental Hazards*, 8(2): 149–63.

Werner, E. E. and Smith, R. S. (1992) *Overcoming the Odds: High Risk Children from Birth to Adulthood*. Ithaca, NY: Cornell University Press.

WHO (World Health Organisation) (2004) *Extreme Weather and Climate Events and Public Health Responses: Europe*. Copenhagen: WHO.

Wiebe, R. and Johnson, Y. (1998) *Stolen Life: The Journey of a Cree Woman*. Toronto: First Vintage Canada.

Wielm, A. (2004) *Digital Nation: Towards an Inclusive Information Society*. Michigan: Michigan Institute of Technology Press.

Williams, C. (2011) 'Geographical Variations in the Nature of Community Engagement: A Total Social Organisation of Labour Approach', *Community Development Journal*, 46(2): 213–28.

Williams, C. and Millington, A. (2004) 'The Diverse and Contested Meanings of Sustainable Development', *Geographical Journal*, 170(2): 99–104.

Winne, M. (2009) *Closing the Food Gap: Resetting the Table in the Land of Plenty*. Uckfield, East Sussex: Beacon Press.

Winter, M. (2003) 'Embeddedness, the New Food Economy and Defensive Localism', *Journal of Rural Studies*, 19(1): 23–32.

Wisner, B., Gaillard, J. C. and Kelman, I. (eds.) (2011) *Handbook of Hazards and Risk Reduction*. London: Routledge.

Wong, L. (1994) 'The Privatization of Social Welfare in Post-Mao China', *Asian Survey*, 34(4): 307–25.

World Bank (2008) *World Development Indicators, 2008*. Washington, DC: World Bank.

World Bank (2010) *World Development Report 2010: Development and Climate Change*. Washington, DC: The International Bank for Reconstruction and Development and World Bank.

World Bank (2011) *Securing the Present, Shaping the Future*. Washington, DC: World Bank.

World Vision (2009) *Raising Resilience: The 2004 Asian Tsunami 5 Years On*. Bangkok: World Vision.

Yaffe, B. (2010) 'Knocking Oilsands Bolsters Northern Gateway', *The Vancouver Sun*, 12 August, p. B2.

Zapf, M. K. (2005) 'The Spiritual Dimension of Person and Environment: Perspectives from Social Work and Traditional Knowledge', *International Social Work*, 48(5): 633–43.

Zapf, M. K. (2009) *Social Work and the Environment, Understanding People and Place*. Toronto: Canadian Scholars' Press.

Zhu, Y. X. and Sim, T. (2009) 'Working with Children in the Sichuan Earthquake Using the "Person-in-the-Environment" Perspective', *Journal of Social Work*, 9(2): 31–4. [In Chinese: 朱雨欣.沈文伟.灾后儿童心理重建路径探析.社会工作，2009年9月下半月，31–34页.]

Some useful websites

www.ecosocialwork.org for 'deep' ecological social work and the Global Alliance for Deep Ecological Social Work.

www.environment-agency.gov.uk for flood plans.

www.ifrc.org for the International Federation of the Red Cross and Red Crescent Societies; provides a considerable amount of information about humanitarian aid, and the crises that they respond to.

www.kildonan.unitingcare.org.au for a discussion of energy audits that involved social workers.

www.med.monash.edu.au/glass for analyses of women and climate change (Global Leadership and Social Sustainability).

www.scie-socialcareonline.org.uk for online information about social work.

www.socialwatch.org for analyses conducted by an NGO; the site includes an interactive map on the impact of the fiscal crisis on all countries in the world for which data are available.

www.socialworkeducation.org.uk for articles on social work.

www.socialworker.com for online articles about social work.

www.un.org for the United Nations website, which is a source of a wide range of information.

www.unicef.org for information on children and the impact of policies on them.

www.uwindsor.ca/criticalsocialwork for online articles o social worker involvement in ecological issues.

www.who.int/hac/network/interagency/news/mental_health_guidelines for the IASC guidelines on gender, mental health and psychosocial interventions in several languages.

www.worldbank.org for information about government policies and actions on development matters.

Author index

Subject index